# HANDBOOK ON HOW TO PREDICT EARTHQUAKES

A *Valid* Theory
Makes *Valid* Predictions

## AHMAD JAMALUDIN

PARTRIDGE

Copyright © 2019 by Ahmad Jamaludin.

| | | |
|---|---|---|
| Library of Congress Control Number: | | 2019937174 |
| ISBN: | Softcover | 978-1-5437-5040-9 |
| | eBook | 978-1-5437-5041-6 |

All rights reserved. No part of this book may be used or reproduced by any means, graphic, electronic, or mechanical, including photocopying, recording, taping or by any information storage retrieval system without the written permission of the author except in the case of brief quotations embodied in critical articles and reviews.

Because of the dynamic nature of the Internet, any web addresses or links contained in this book may have changed since publication and may no longer be valid. The views expressed in this work are solely those of the author and do not necessarily reflect the views of the publisher, and the publisher hereby disclaims any responsibility for them.

Print information available on the last page.

To order additional copies of this book, contact
Toll Free 800 101 2657 (Singapore)
Toll Free 1 800 81 7340 (Malaysia)
orders.singapore@partridgepublishing.com

www.partridgepublishing.com/singapore

# CONTENTS

Introduction ..................................................................... vii

| | | |
|---|---|---|
| Step 1 | Find The Root Cause ........................................... 1 | |
| Step 2 | Consider This—If It Is Not From Below, Could It Be From Above? ............................................. 4 | |
| Step 3 | Find The Culprit (Stimulus) In The Sky ............ 7 | |
| Step 4 | Understand The Observed Behaviour Of Earthquakes; Then You Will Find The Culprit! ............... 10 | |
| Step 5 | Correlate The Presence Of A Hypothetical Source X To Explain The Behaviour Of Earthquakes ...... 12 | |
| Step 6 | Observe How Earthquakes Are Distributed Geographically ................................................ 21 | |
| Step 7 | Check The Behaviour Of The Orientations Of The Straight Line Alignments (Slas) ................ 29 | |
| Step 8 | The Seesaw Effect Of Earthquake Epicentre Distribution ...................................... 47 | |
| Step 9 | Determine The Shift And Flip Angle Of Source X ......... 52 | |
| Step 10 | See How And Why Large-Magnitude Earthquakes Happen ............................................ 57 | |
| Step 11 | Preliminary Protocol For The Prediction Of Earthquakes ........................................................ 76 | |
| Step 12 | Crosscheck The Validity Of The Prediction Protocol ...... 86 | |

References ........................................................................ 101
Appendix A ..................................................................... 103

# INTRODUCTION

Can earthquakes be predicted? The general consensus is that they cannot. This is due to the fact that there are several factors and unknown variables in play in each specific geographical location. Any observation of earthquake precursors, such as foreshocks, changes in geomagnetic and gravity fields, radon counts, anomalous animal behaviour, made before an earthquake in one location cannot be used in another area to predict one. A good example is the 1975 Haicheng earthquake in China. Its successful prediction, however, cannot be duplicated elsewhere.

Each earthquake in a given area seems to be dictated based on the local geophysical conditions at the time. If this notion is true, then no single formula can be used to forecast all global earthquakes.

Any valid earthquake prediction must satisfy three important elements: the location of the quake, the time of occurrence, and its magnitude. If we can also predict its depth, that would be an added bonus.

All these can only be achieved if we are able to identify the primary cause or triggering mechanism of earthquakes. However, at present we can only see its effect, with the possible cause or the actual stimulus still in obscurity.

To make any valid earthquake prediction, the theory to its cause and effect must be correct. Towards this, we need to have a radical new theory. If there is a common denominator for all earthquakes (with the exception due to mining activity, quarry blasts, and nuclear testing), then seismology will have a brighter future.

The three keywords mentioned repeatedly in this book relate to seismology: plate tectonics, fault line, and epicentre. Plate tectonics refers to the structure of the earth's crust and the interaction of the plates that moved over the underlying mantle. A fault line is the surface feature with a break or fracture, where an earthquake occurs. Epicentre refers to the point on the ground surface above the focus point in the crust where the seismic rupture took place.

Also mentioned here to accommodate the theory of an extraterrestrial stimulus of earthquakes are terms relating to astrophysics. An astrophysical object refers to compact stellar remnants such as white dwarfs, neutron stars, and black holes. The theoretical wormhole is also considered as an astrophysical object.

Gravity waves, or gravitational waves, carry energy in the form of gravitational radiation. Sources of gravity waves emanate or originate from compact masses such as neutron stars and black holes. There could also be other astrophysical objects or compact masses of energy that we have not yet discovered or identified.

An astrophysical object does not necessarily imply a solid object. It could also be a compact mass of energy. The hypothetical source mentioned in this book as the culprit in triggering earthquakes refers to this type of astrophysical mass. This compact mass is the source of the gravity waves or energy that is believed to bombard the Earth.

The term *beam sweep* refers to this type of gravity wave, whereby its energy beam draws an imaginary straight line on the face of the Earth. We have coined the abbreviations such as straight line alignment (SLA) and straight line corridor (SLC), which actually refer to these energy beam sweeps. SLA can tell us where the quake will strike, whereas SLC will tell us how big the quake will be.

By combining the field of seismology and astrophysics, we see a chemistry that relates the trigger source and earthquakes in a perfect combination to formulate a prediction protocol or a foundation of where and when to expect an earthquake.

To use an analogy, if we see only bullet holes on the wall, there must be a way to determine what is causing the holes and where is it coming from. If we are registering only the earthquake epicentres, there must be a mechanism or foundation to identify what is causing the quakes and its origin.

The basic framework of earthquake mechanics is based on an entirely different perspective than is proposed here. The observed pattern outlined allows us to make only a general prediction of where and when the tremor will occur.

At this early stage, there is no necessity to complicate the issue by trying to predict the magnitude and the depth of the quake as well. Let us take one step at a time. We need to confirm that the theory presented and the observations and findings made in this book are correct.

From there, seismologists and astrophysicists will need to refine the findings to construct a more precise model and its mathematical calculations to warrant a successful earthquake prediction.

The Wright brothers gave us the basic framework on how to build a flyable machine. This handbook hopefully provides a similar basic foundation to understand how earthquakes occur and how to successfully predict them in time and space.

Step 1

# FIND THE ROOT CAUSE

Any medical problems relating to the human body are due to external or internal factors. Externally, they could be caused by pathogens or carcinogens. Internally, they could be due to the breakdown of body metabolism relating to the production of hormones or enzymes.

Early in the history of medicine, no one knew that external agents could cause diseases. The notion at that time was that they were all internal problems. With the discovery of bacteria, our perspectives changed.

The same could be said of the situation facing Earth regarding earthquakes. Externally, the Earth is bombarded by cosmic radiation, magnetic storms, neutrinos, gravity waves, and so on. Internally, they could be due to the breakdown of its ecosystem, weather, geologic stress, and more. So where do earthquakes fit in? Are they all due to some internal geological processes, or is there an external stimulus in play?

In the early 1930s, several theories were put forward to explain plate motion and continental drifts in the Earth's mantle. It was initially attributed to large-scale convection currents. Later theories added further dynamics involving friction and gravity. This was how it was believed that earthquakes occurred.

The accepted theory, at present, is that earthquakes occur due to:
1. Plate tectonics, or large-scale motion of the large plates and the movements of a larger number of smaller plates of the earth's crust. Plate collision and sliding can result in earthquakes.
2. Elastic rebound theory, which explains how energy is released during an earthquake. As adjoining plates on the Earth's surface move in opposite directions, the rocks that span the opposing sides of the incipient fault are subjected to extreme shear stress.

The above theories, however, do not necessarily constitute the actual trigger for earthquakes to happen. They only speak of plate motion or energy release. But what actually set them in motion in the first place? What caused the associated effects?

The original idea, posed by Alfred Wegener in 1929, believed that convection currents in the Earth's mantle are the major driving force of plate tectonics.

If the theories presented over the decades on the dynamics of earthquakes are correct, we should be able to predict earthquakes with some degree of accuracy by now. Unfortunately, each day we are still caught dumbfounded. Mega earthquakes continue to occur regardless of what the theories envisage, and we are still left clueless to forecast the next quake in any given area.

Perhaps it is time that we look for a new theory from an entirely different perspective. Over the years, there have been no concrete observations to aid in the forecasting of earthquakes. Although the theory of plate tectonics is appealing to all, its outcome does not and cannot indicate where and when a quake will strike at any given area and time.

The question we should be asking, for a bigger picture on a universal scale, is, "What actually triggers earthquakes, moonquakes, or even sunquakes (which could be manifested in the form of sunspots, where an outside external gravity neutralises the sun's gravity, resulting in the formation of cool spots)?" At present, our understandings of sunquakes are in the form of giant outbursts of magnetic energy originating above the sun's surface.

Understandably, quakes must also be present on Mars and all the planets in this solar system. If all the seismic upheavals on the planets are not of local origins, the stimuli must be coming from somewhere else, possibly external in nature.

The theory of plate tectonics cannot explain moonquakes because the moon is a more compact mass when compared to the Earth. It does not have the mantle fluidity as on Earth. It certainly does not work with the sun because it has no solid ground plates. Its texture and violent physical nature do not support any plate movements. Jupiter, being a gas giant, does not quake, but it too must feel some form of interference. This could manifest in the form of wobbles instead.

To make any successful forecast on earthquakes, we have to look for another answer. And that answer must probably be something new to explain what is happening on earth and all the planets in this solar system.

Because we cannot find the answer in the ground, why not look for it in the sky?

Step 2

# CONSIDER THIS—IF IT IS NOT FROM BELOW, COULD IT BE FROM ABOVE?

Because we could be dealing with a new unidentified subject, in this handbook we need to put the cart before the horse. This is unorthodox, but what if the source or powerhouse of our earthquakes is not solely from the ground but triggered by an extraterrestrial source?

We have always looked in the ground to find an answer to our earthquake problem. The answer that we have been looking for is not forthcoming, even with the aid of modern instrumentation and computer analysis. But what if the answer is not really in the ground but in the sky? In their book *The Jupiter Effect*, astrophysicists John Gribbin and Stephan Plagemann envisaged that the gravity of Jupiter can induce earthquakes on Earth. They theorised that an alignment of the major planets in March 1982 on the same side of the sun would trigger a series of cosmic events and a devastating earthquake that would wipe out Los Angeles. However, this did not happen.

Global earthquakes, minor and major, occur on a daily basis. No major planetary alignment can take credit for that. On the whole, Jupiter would exert very little effect. The gravitational force of Jupiter would act en masse on Earth or any other nearby planets.

However, the statistics and behaviours of earthquakes on Earth suggest otherwise. Earth seems to be surgically bombarded regularly by something sinister that's from not within the earth but possibly from an external stimulus. In other words, the trigger force could possibly come from a source that could be pulsating with gravity waves or energy. This would be the most probable effect to cause the thousands of earth tremors each year in a pinpointed nature.

To make any successful earthquake prediction, we need a new approach. This handbook proposes a theory that all the quakes in this solar system originate or are triggered by a single astrophysical source (or a kind of powerhouse in the heavens).

Where this triggering source is located is a matter for future discussion. If this theory is valid, then all earthquake observations will fall in place. That is, earthquakes occur one after the other in an orderly fashion.

We call this triggering point in the sky Source X. Source X constitutes the powerhouse that ejects gravity waves. When these waves strike the earth surgically or locally, they cause offsets of the Earth's gravitational grip at the fault lines, causing earthquakes.

To understand how and why earthquakes occur, and the possible role played by Source X, let us take a look at the following analogy. Imagine you are in a cosy, warm room, and suddenly a gust of the cold air seeps through a crack in the wall. The initial entry of the cold air does not offset the temperature of the room; you do not feel the difference. But as the process repeats itself several times, the room finally becomes cold until it reaches an unbearable condition, and you begin to shiver.

The same scenario seems to occur with earthquakes. Imagine that you are standing on stable ground, and say that an unknown form of energy from the heavens suddenly bursts into the atmosphere and enters through the cracks in the ground at the fault lines. As we know, the earth is held together by gravity. We will not feel any tremors during the initial strike.

As this energy crosses the area several times, it builds the stress and pressure in the ground to destabilise the Earth's gravitational grip, just as the cold air overcomes the warm air in the room. If we shiver with the cold air, the Earth trembles or quakes as gravity momentarily loses its equilibrium.

With this new theory of Source X, we should be able to explain the how and why of earthquake occurrences from an entirely different perspective.

Step 3

# FIND THE CULPRIT (STIMULUS) IN THE SKY

The heavens are filled with energetic and exotic astrophysical objects, some of which we have discovered. Perhaps many more are still hidden in the depth of space. Among those that have been recognised are neutron stars (or pulsars), quasars, and black holes. Some of these objects have massive gravity levels and could have a major influence on the well-being of other planets and stars in their vicinity.

If such an object existed nearby, we could expect that life on Earth would not to be a smooth one. If we take earthquakes, for example, they have caused serious damage to property and lives. Hundreds of tremors occur each day, and sometimes we are jolted with a big quake. Those in earthquake-prone areas cannot get a good night's sleep.

Earthquake mitigation, or hazard programs, cannot be put into place effectively if we do not know where and when an earthquake is going to strike. To get everything in order, we first need to identify the culprit or the actual trigger factor that causes earthquakes.

The hundreds of studies made over the years relating to seismic ground dynamics—such as foreshocks, radon counts in wells, and magnetic and gravity anomalies—have failed to identify the precise indicators of an impending earthquake. Perhaps the search for the ground indicators works better if we can correlate it with something in the sky.

If earthquakes were triggered by an extrasolar coordinate in the sky, we should expect that there are also quakes in the other planets in this solar system. For example, the tiny planet of Mercury, with its active seismic hills, is the only known planet in the solar system that is tectonically active like earth. Mars may have marsquakes because the evidence shows it has tectonic boundaries.

The bottom line is if everyone in the neighbourhood is having the same type of flu, then the causative agent must be coming from somewhere outside.

Are our earthquakes the direct result of the presence of such a menacing astrophysical source? Is there a possibility that such a source existed nearby, right in our own back yard, which has remained undetected all these time?

Assuming for a moment that there exists an astrophysical powerhouse with the right type of energy and dynamics in this solar system, it could theoretically induce or trigger earthquakes not only on Earth but on all the planets in this solar system, the sun included.

How do we show or prove that such a powerhouse has existed all along, unnoticed and undetected for such a long time? If such an object is present, we should be able to see a pattern in the distribution of earthquakes in space and time. But to see the pattern, we need to correlate with the possible characteristics or behaviour of the mystery source, if it is really out there.

Theoretically, how would this source be behaving, and what characteristics does it have?

To get an idea to the nature of Source X, we assumed that its basic properties (like most energetic astrophysical objects) must include spinning or rotating and has a certain angle of tilt and emits some form of radiation, gas, or energy emission. These energy emissions may be in the form of pulses or a steady stream of energy, like the light coming from a lighthouse.

In theory, Source X would exhibit the following hypothetical characteristics:
1. It is a small object with two poles. The poles eject gravity waves.
2. These waves or energy beams are very small in width and are very directional.

3. It rotates on its axis, thus resulting in the energy beam striking Earth in short duration and in many criss-crossing beam sweeps.
4. It has three speeds: normal (which is slow), intermediate, and fast when it is energetic.
5. At slow speed, Source X would wobble, resulting with the criss-crossing beam lines not intersecting at a common point on Earth.
6. When its rotation speeds up or glitches, it is more stable, the beam sweep will be directed to a small geographical area, and we can see many lines crossed at a common point.

Let us correlate the behaviour of the hypothetical Source X with the observed characteristics of earthquakes in time and space, as outlined in the next chapter, and see whether it fits the observations.

## Step 4

# UNDERSTAND THE OBSERVED BEHAVIOUR OF EARTHQUAKES; THEN YOU WILL FIND THE CULPRIT!

Geographically wide apart earthquakes appeared to be random events unrelated to each other. That is, the quakes in Japan have nothing to do with the quakes in Chile. Each earthquake is being treated as a local event due to local stress and tectonic plate slides, abrasions, or collisions. If a series of them occurred close together, it is attributed to sharing the same plate boundaries.

The accepted theory is that plate movement is due to mantle dynamics, such as its viscosity, convection currents, and heat dissipation, which is believe to be the driving force to move the plates. Gravity too played its part.

Remember that a coin has two sides. If we constantly looked at just one side, in time we will hold to the idea that both sides are the same, or we have forgotten that it has its other side too.

Now, the general acceptance is that plate tectonics caused earthquakes, and the primary focus has been to this side only. What about the other side of the coin? Is there a possibility that the other side will give us an entirely different perspective of how and why earthquakes occur?

This book flipped the coin to the other side, placing earthquakes at a different angle. And this angle attributed the cause of earthquakes to something in the sky.

Let us observe the modus operandi of earthquakes, which showed us the following characteristics:

1. Earthquakes strike at a specific point on the ground, known as epicentres. Its actual point in the ground is called the hypocentre.
2. Earthquakes appeared to be randomly distributed in time and space. But at times, they are concentrated or clustered in a small geographical region.
3. Tremors occurred in varying intensity, ranging from microquakes to megaquakes measuring 8.0 or greater on the Richter scale.
4. Large quakes are usually accompanied by many smaller magnitude tremors known as aftershocks.
5. There are times when earth tremors comes in the form of earthquake swarms. These are in the form of low-magnitude tremors in a confined geographical area.
6. Several quakes in a given time frame appeared to be arranged in a straight line alignment across the globe.
7. An earthquakes is of short duration. It is transient in nature. It does not shake continuously for hours or days.
8. It is not possible (or it's rare) to find two earthquakes occurring at precisely the same time across the world. It is like a person with a single gun cannot shoot two different targets wide apart at the same time.
9. Generally speaking, small quakes occur every minutes, but megaquakes come only once in a year or more. It is like when throwing a dart, if you are not a pro, most of the darts will miss the bull's-eye. But occasionally you get it right on target. Something is missing its mark to cause a major quake. A wobble or misorientation can cause this.

As the saying goes, study the modus operandi of a thief in order to catch a thief. For seismic events not relating to mining activities and quarry blasts, we need to study the behaviour or characteristics of earthquakes to catch not only the next impending quake but also the primary culprit behind the ground upheavals.

So what could possibly be causing or triggering the above observations?

Step 5

# CORRELATE THE PRESENCE OF A HYPOTHETICAL SOURCE X TO EXPLAIN THE BEHAVIOUR OF EARTHQUAKES

Imagine that if someone has not seen a ghost, and it has not been scientifically proven to exist, then how do we show that ghosts actually exist? We cannot, unless we can come out with some observations to support the existence of ghosts. To do this, we need to correlate its time of appearance, its location, and its manifestation characteristics. If the findings followed a certain set of patterns, then we need to correlate it with our brains and psychological patterns to see if it fits.

If not, then the manifestation of ghosts is not in our minds. It has to be outside of our perspective of reality. They are not actually real in the sense that you can touch and feel them, but they are real in the sense that your physical vision can detect them even though ghosts are not physical entities.

To convince others that an earthquake is triggered by a source that is not visible to the naked eye, but that its effect is visible, is like trying to show that ghosts exist. As with earthquakes, we have all the data pertaining to its location, time, and seismological characteristics. What we need to do is show that there are pertinent patterns in the distribution and occurrence of earthquakes in relation to an unknown invisible source.

At present, the seismological community does not see any empirically convincing patterns in earthquake distribution in time and space. A paper cites a doughnut-shaped pattern in seismic activity in 1978 at Shimane, Japan. This trend was also seen in the Wenchuan seismic region in China. These observations, however, do not help in making seismic predictions. They simply show a pattern on how Earth tremors occurred collectively in a region.

However, this trend or any other observation (such as the occurrence of foreshocks, earthquake lights, and abnormal animal behaviour) usually does not repeat in every seismically active area. That is why we are unable to predict or forecast earthquakes for the past one hundred years.

This book hopes to throw some light on a new and different perspective: that there is a "ghost" that excites earthquakes. Find this ghost, and your understanding of earthquakes will change for the better.

But before that, we must show that some important and undeniable patterns in the distribution of earthquakes do exist. The pattern that we assumed inherited from Source X must be reflected in the distribution pattern of earthquakes in time and space on Earth.

A ghost does not leave any footprints. But Source X, being a physically active powerhouse, will leave many footprints as its energy beam criss-crossed the face of the earth each day. The footprints of Source X were the Earth tremors. It excites the fault line when it crossed them. Follow these footprints, and we can guess where the source is.

Let us now examine whether the hypothetical behaviour or properties of Source X corresponded to the nature of earthquakes that we are experiencing. Let us answer the questions about earthquakes with respect to this hypothetical Source X.

## 1. Why Do Earthquakes Occur at Fault Lines and of Varying Magnitudes?

As the theory postulates, Source X rotates on its axis, and the energy that came out from its two poles will sweep across the face of the Earth like a searchlight beam. Unlike a searchlight beam, which is small at its apex and widens with distance, the beam from Source X is a constant small diameter from the source to the target. Imagine it like an infrared pointer, whereby it paints a red dot on the target.

With this in mind, imagine it as a narrow and directional energy beam passing over the Earth. We can draw the ground track of the energy beam as an imaginary straight line, A to C (refer to Figure 1 below).

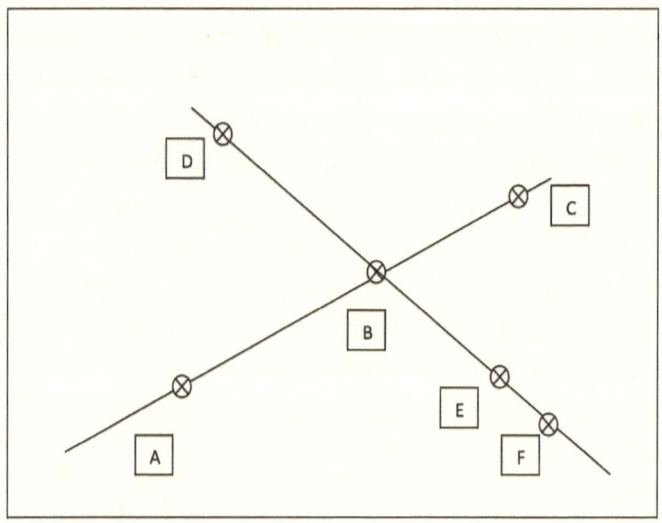

*Figure 1. Ground track of hypothetical beam sweep on Earth.*

As an example, as the beam sweeps past an area, it will pass over locations A, then B, then C, and so forth. All these locales had fault lines intersected by the straight line. This first beam sweep will exert a stress on all the fault lines that it crossed. We will call the induced stress *p1*. But all the locations will not endure the same amount of stress because the geophysical condition at each fault line varies.

As Source X makes a complete 360-degree rotation, the next beam will make the new sweep. During this second sweep, due to its changing speed, it will probably wobble and resulted in a tilt in its angle of rotation. When this happens, it will not follow the same path of A to C and so on.

Assuming the new path is D, and then E and F, we see that it crosses location B for the second time. The stress at location B has become *p1* + *p2*. Depending on the geophysical condition of locale B, the stress could induce either a minor quake or a higher magnitude tremor.

This is how an earthquake can occur in a given area. To sum up, this is what happened:

1. The stress in the ground built up as the beam swept over the area two or more times.
2. The magnitude of the quake is directly proportional to the number of beam sweeps over the same area and the geophysical condition of the ground.
3. As a simple guide, *p1* will cause a minor quake, *p1* + *p2* will cause a major quake, and *p1* + *p2* + *p3* or more will result in a mega shock.

## 2. Why Do Major Quakes Come with Aftershocks?

The present understanding is that aftershocks are the direct result of a major quake. But if we look at this new theory, major shock and aftershocks are not directly related. In other words, a major earthquake does not produce numerous aftershocks. Aftershocks occurred on their own merit.

Megaquakes occur as a result of the beam sweep that stabilises and zeroes in on a small specific area. When this happens, the lines or the ground track of the beam sweep will crossed a specific foci several times. The stress induced at the foci will be *p1* + *p2* + *p3* + *p4* and more. The stress built up will manifest as a major quake.

Meanwhile, areas surrounding the foci will also received a considerable amount of criss-crossing of the beam sweep but less than at the foci. Because the surrounding nearby areas of the main shock received a lesser degree of stress of less than *p4* or so, it will be registered as tremors of a lesser magnitude.

We can say that aftershocks are collateral events for being in the area and line of a major earthquake. In other words, aftershocks are the effect of being in line near the foci of the major shock.

Any major quake will occur at the point where several beam or straight line alignments crossed at a common centre, as depicted in Figure 2 below.

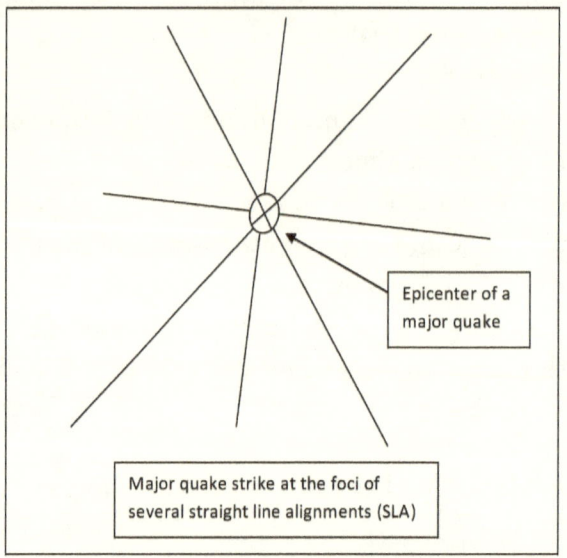

*Figure 2: The spoke wheel effect, which can cause a major quake.*

To sum up, we can see the following in a major quake zone:
1. The imaginary lines connecting all the epicentres surrounding a major quake will show wheel-like spokes as most of the beam swept across a common centre (refer to Figure 2). This point is the epicentre of the major quake because it received the most stress.
2. For a major quake to occur in any area, Source X must speed up its rate of rotation. This will put it in a stable position. When this happens, its beam sweep can be focused into one specific area. It is like if we place a shooter on a shaking table, the shots will hit all across the wall. But if the table remains steady, the shooter can aim directly to the intended target.
3. When Source X loses its momentum of rapid rotation, its speed decelerates. The low speed will cause it to wobble. It is like a spinning top. When at the top speed of rotation, its base

remained steady on one spot. But when it loses speed, it begins to wobble, and its base is described as an erratic circle. When this happens to Source X, its sweeping beam will widen again and criss-cross the face of the Earth with isolated, random quakes.

## 4. Why Do Earthquakes Strike at Different Depths?

Image you hammered a nail to a wall. At first strike, it will penetrate the brick a notch. You gave it a second strike, and it will go in another notch. How many strikes it takes to get the nail into the brick will depend on the force used, the nature of the brick, and how many times it took you to strike.

The recorded depth of an earthquake is similar in nature. It will depend on the energy striking the fault line, how many times it passes over the area, and at what depth the critical limit of the rocks can endure before it finally breaks down due to the stress. By this notion, we can understand why earthquakes strike at various depths.

## 5. Why Are Global Earthquake Epicentres Aligned on Straight Line Corridors?

If you observe the spatial distribution of global earthquake epicentres on any given day, you will notice that there are several epicentres located on a perfectly straight line across the continents. These epicentres' arrangement appear like collinear points. This cannot be a coincidence because it repeats itself but with changing compass orientations, continuously and on a daily basis.

The straight line alignment of epicentres on a map does not represent the true picture of earthquake mechanics because the Earth is a globe. However, these alignments are on the great circle lines. This shows that though earthquakes are internal events, the real stimulus could probably be external or extraterrestrial in nature.

In 1956, G. MacCarthy, from the Department of Geology and Geography, University of North Carolina, observed that the earthquake data from 1885 till 1956 showed that at least eleven earthquakes were aligned in an almost perfect straight line in a north-eastern direction across western North Carolina and up into Virginia. In fact, if we observed all close-proximity quakes in time and space, they are all aligned in a general north-east or north-west orientations.

As an analogy, imagine a high-flying stealth bomber flying on a straight course. It drops a single bomb every one minute. If you plot the bomb craters, you will see that it is arranged on a general straight line. The same is true if the trigger mechanism of earthquakes is coming from the sky. If we plot the epicentres, it should be aligned in a straight line.

So why are earthquake epicentres aligned in a straight line? This is the result of the beam sweep, whose path is a straight line. Please refer to the chart below.

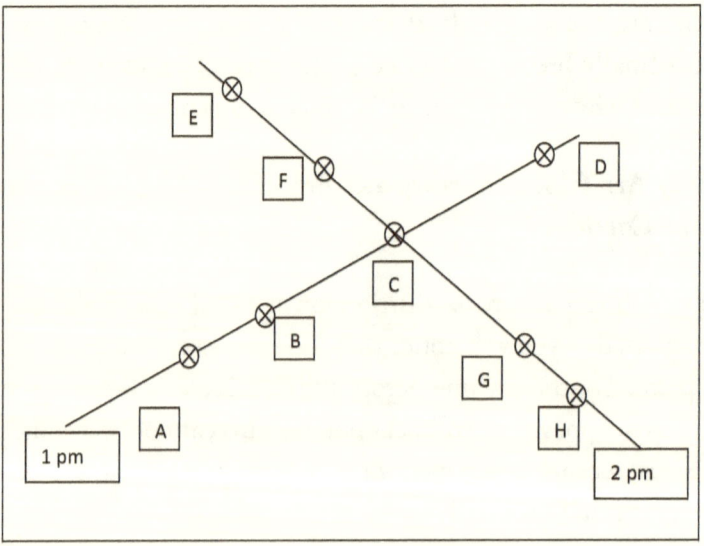

*Figure 3. Ground track of hypothetical beam sweep on Earth.*

Say an energy beam swept across the face of the Earth at 1 p.m. It passes over locations A, B, C, and D. All these locations will be aligned in a straight line. At 2 p.m., the energy beam made another sweep but from a different compass orientation. Now we have a second straight line of E, F, G, and H.

Something interesting to note is that in this example, the second beam sweep crossed location C for the second time. If location C was a fault line, the second pass could induce a considerable amount of geological stress to it. A tremor could occur.

In reality, this is what is happening on Earth. The constant beam sweep of gravity waves excited the fault lines, which manifested themselves in the form of an earthquake. The straight line orientation of the earthquake epicentres actually reflects the straight line nature of the energy beam sweeps.

## 6. Why Do We Have Earthquake Swarms?

Earthquake swarms are believed to be the result of the intrusion of fluids in the seismic area. The mechanics behind it, however, are still not fully understood.

Unlike the major quakes, which have a series of tremors accompanying the main shock, earthquake swarms lack a major event. It is simply a cluster of minor tremors in a confined geological region.

To better understand earthquake swarms, there must be a clear distinction to separate or define an earthquake swarm from the numerous aftershocks following a major seismic event. Earthquake swarms are of very short duration. They cannot last continuously for months but could repeat at a later date.

Earthquake swarms, as an analogy, are similar in nature to the sensation of being pricked by hundreds of needles—or in our case, being bombarded with numerous low forms of energy. The only difference is that the prickling sensation in humans may be due to some biomedical process, or some are attributed to the close proximity to electronic apparatus. It could be either internal or external.

At present, the theory regarding earthquake swarms is believed to be of internal origin. The presence of a hypothetical Source X can gives an external explanation. Which one would give a better picture?

For the Source X theory, the energy beam sweeping over the area cannot explain the events. The only possibility for an earthquake swarm to happen is that Source X has to stop rotating for a short duration!

When this happens, its beam is directed to only one small region. However, no major quake will occur during this time because it has lost its maximum energy burst. That is why it has stopped spinning in the first place. When fresh energy is again ejected out, it will pick up its momentum, and the rotation will build up speed again. Therefore the thousands of low-magnitude energy beams will cease.

Step 6

# OBSERVE HOW EARTHQUAKES ARE DISTRIBUTED GEOGRAPHICALLY

Most seismologists seem to pay attention to a specific geographical area of their immediate concern or interest. If earthquakes shared a common denominator, we could miss the global spatial-temporal pattern altogether. Any attempt to plot or chart all the global earthquake epicentres in days or months will not show any pattern on the map. This is because the numerous epicentres will mask any discernible trend.

Let us follow the distribution of global earthquake epicentres for any given day. In this example, we have chosen 29 April 2017. The data is from the USGS Earthquake Catalog. To get an accurate picture, we must use all the earthquake data from magnitude 1.0 to the highest recorded for the day.

The USGS Catalog lists earthquakes of greater than magnitude 1.0 for the US events only. Other global quakes of less than 2.4 magnitude are not listed.

Based on this insufficient data, we can still try to build up a general picture of how the global earthquake epicentres are distributed in time and space.

Table 1 below list some of the earthquake epicentres recorded on 29 April 2017, which appears to have a straight line alignment (SLA). When we include earthquakes epicentres along this line in a five-kilometre width, it will be called a straight line corridor (SLC).

The epicentre locations in this book were identified by their time of occurrence (decimals omitted) and not their geographical coordinates.

The straight line is measured from quake A to quake C or D (depending on whether there are three or more epicentres). The deviations are measured in kilometres from the line connecting point A to C or D. The deviation refers to the epicentres for quake B, C, or more. Google Earth Pro is used to measure the deviation distance from the straight line, the length of the straight line, and its compass orientation.

The maximum width of the corridor is taken at five kilometres. This is considered to be a reasonable one because any directional beam of energy coming from outside of Earth cannot give a perfect straight line on the second or succeeding passes. It is like if someone made a straight line on a wall, the later second straight line will not pass directly above the first line. This could be due to the stability factor induced by the concentration of the hand and mind. The same situation is expected for any rotating extraterrestrial beam that will not make a straight line exactly on the first line that it made earlier. This could be attributed to its stability or wobble factor.

However, there is a chance that we can see a perfect straight line if the source stabilises for a certain time frame. This happened during an eleven-minute time frame whereby three epicentres are on a perfect straight line. This line spanned over eight thousand kilometres long with earthquakes recorded at 000116Z (near Cordova, Alaska), at 001256Z (the British Virgin Islands), and at 001306Z (near Old Iliamna, Alaska).

Because these three epicentres did not occur in an orderly fashion, we can assume that the 001306 tremor occurred on its second pass or succeeding passes. Figure 1 below demonstrates the perfect straight line between Alaska and the British Virgin Islands.

The straight line distribution of earthquake epicentres for 29 April 2017, events are given in Table 1 below.

Table 1: Earthquake Epicentres on Several Straight Line Corridors on 29 April 2017

| Quake A | Quake B | Quake C | Quake D | Quake E | Dev. for B (km) | Dev. for C (km) | Dev. for D (km) | Dist. (km) | Orientation (Deg.) |
|---|---|---|---|---|---|---|---|---|---|
| 095858 Oklahoma | 094815 Oklahoma | 034815 Kansas | 222552 Kansas | 035703 Kansas | 0.02 | 4.15 | 3.19 | 61 | 343 |
| 151424 P. Rico | 062101 Wyoming | 021117 Montana | — | — | 2.65 | — | — | 5134 | 315.7 |
| 181920 California | 210358 Utah | 191841 Utah | 175128 Utah | — | 5.21 | 4.18 | — | 737 | 35 |
| 104042 California | 040005 California | 191818 California | — | — | 0.33 | — | — | 310 | 83.6 |
| 165058 Taiwan | 224141 Japan | 202259 Japan | — | — | 3.18 | — | — | 1481 | 42 |
| 224119 Japan | 224141 Japan | 234917 Mariana | — | — | 3.74 | — | — | 2335 | 9.7 |
| 183923 Indonesia | 131630 Philippines | 083625 Japan | — | — | 4.31 | — | — | 2620 | 4.9 |
| 231522 Vanuatu | 133546 Solomon | 225930 Philippines | — | — | 3.93 | — | — | 5150 | 300 |

| | | | | | | | | | |
|---|---|---|---|---|---|---|---|---|---|
| 173733 Indonesia | 100212 Indonesia | 143004 Indonesia | 165028 Indonesia | — | 1.42 | 4.72 | — | 2347 | 51.6 |
| 234917 Mariana | 194049 Japan | 123234 Japan | — | — | 2.37 | — | — | 1966 | 316 |
| 114250 Alaska | 080134 Tonga | 085013 Tonga | — | — | 4.75 | — | — | 8214 | 3.9 |
| 114250 Alaska | 015018 Hawaii | 090213 Hawaii | 222039 Hawaii | — | 3.88 | 4.50 | — | 4012 | 348 |
| 081534 California | 135155 California | 045539 California | — | — | 0.55 | — | — | 160 | 75.8 |
| 210358 Utah | 191841 Utah | 175128 Utah | — | — | 1.48 | — | — | 295 | 35.9 |
| 210358 Utah | 142411 Nevada | 204359 Nevada | — | — | 0.06 | — | — | 460 | 337.6 |
| 081247 California | 024713 California | 024954 California | 024716 Oregon | — | 3.02 | 0.95 | — | 378 | 32.6 |
| 015339 California | 183518 California | 212212 California | 011940 California | 134118 California | 0.56 | 1.92 | 0.44 | 162 | 315.7 |
| 040005 California | 095918 California | 040446 Alaska | — | — | 2.31 | — | — | 3429 | 334.5 |

This trend can be seen every day for all earthquake data. Earthquakes therefore seem to adhere to a predetermined rule. If we follow this rule, it would be possible to determine where the next quake will occur.

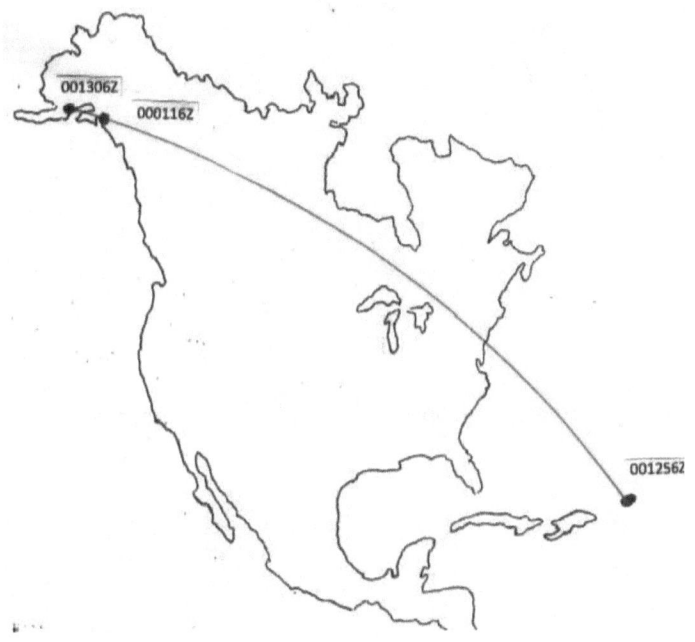

*Figure 4: A perfect straight line for three epicentres between 000116Z–001654Z, from Alaska to the British Virgin Islands.*

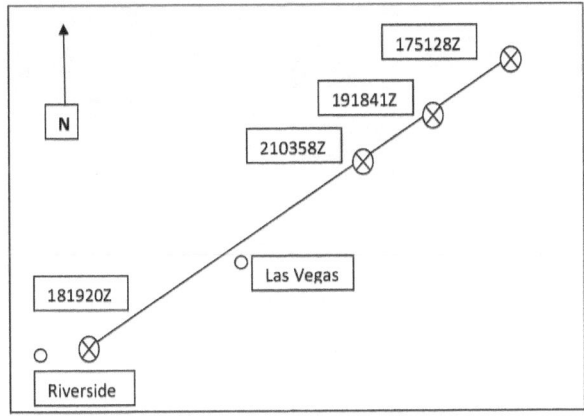

*Figure 5. The Straight Line Corridor for a 4-Point Epicenter in the US from California to Utah*

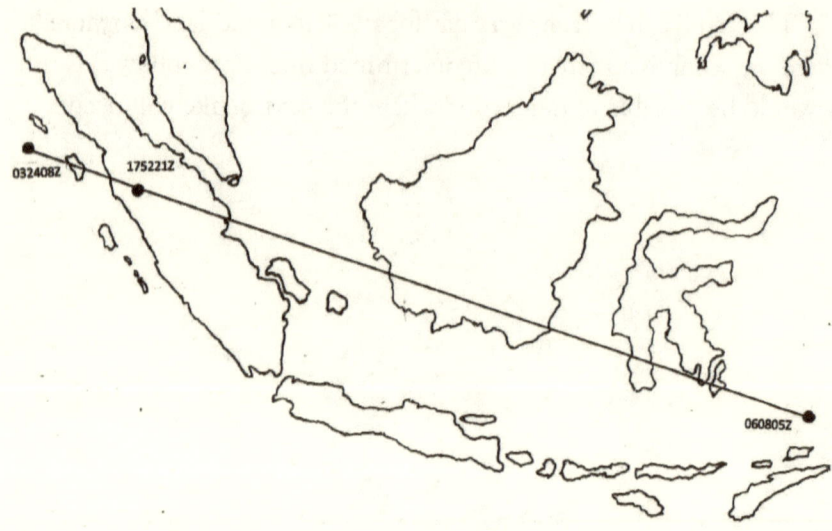

*Figure 6. The straight line corridor for a three-point epicentre in Indonesia.*

*Figure 7. The straight line corridor for a three-point epicentre near Taiwan to Japan.*

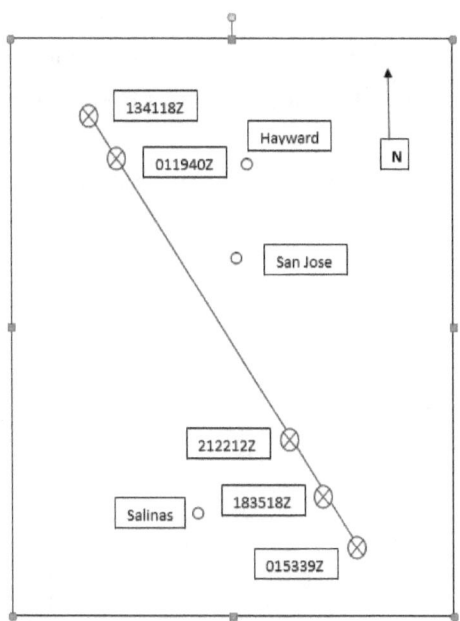

*Figure 8: The straight line corridor for a five-point epicentre in California.*

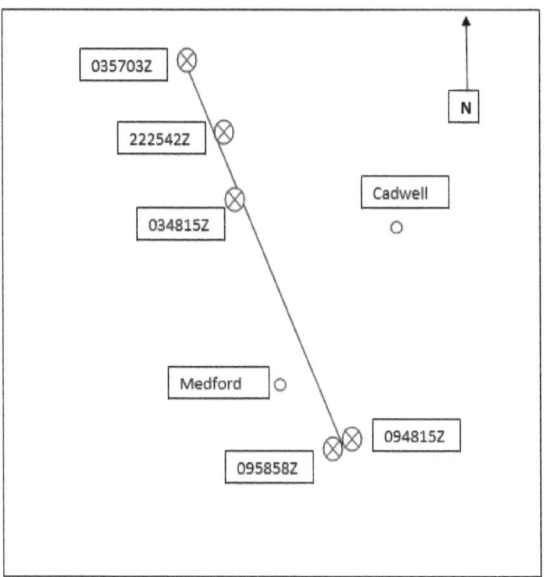

*Figure 9. Five earthquake epicentres in a 61-kilometre-long straight line corridor 5 kilometre wide from Oklahoma into Kansas.*

To make any successful earthquake prediction, we need a certain pattern to start with. This straight line alignment is a good first step. However, the straight lines alone do not provide enough pieces to the jigsaw puzzle.

The next thing to do is look for a pattern on how the straight lines behave in time and space.

Step 7

# CHECK THE BEHAVIOUR OF THE ORIENTATIONS OF THE STRAIGHT LINE ALIGNMENTS (SLAS)

To get a clear picture on the geographical orientation of earthquakes with respect to true north, we need to isolate each individual orientation line. Two consecutive earthquakes that occurred one after the other, regardless of their geographical locations, is taken as its shared line of orientation. Each of the lines, with respect to the source, should be representative of the beam sweep across the face of the globe during that given time frame.

The lines below represent two tremors occurring in succession, one after another. The orientation is in reference to true north. The circle represents the Earth. A straight line refers to two epicentres on the same alignment (SLA). This line is in fact the track of the beam sweep.

The line, however, is not the actual line when the earthquake occurred. It is just the footprints of the first beam sweep that induced the stress *p1* to all of the two foci on each individual straight line.

This approach can be contradicted by saying that any two locations on Earth will form an alignment. But unrelated location for any events on an alignment is meaningless unless the events share a common trigger. Take, for example, if we plot two alignments for armed robberies consecutively occurring worldwide. The line alignment has no relationship although the crimes are the same, but the culprits are of different personalities.

But what if earthquakes were triggered by the same source? Regardless of in which part of the world the quake occurred, each quake is directly related with each other. In other words, one line alignment will determine the orientation of the next alignment. This means that each single quake will determine where the next one will occur.

The line alignments can be used as indicators to see the progression of the tremors in time and space. With this in mind, we can guess where the next quake would be.

To get a picture of how earthquakes are distributed globally in time and space, let us follow the progression of the straight line alignments for a 2-hour, 47-minutes time frame on 29 April 2017, for the period 000116Z to 024731Z.

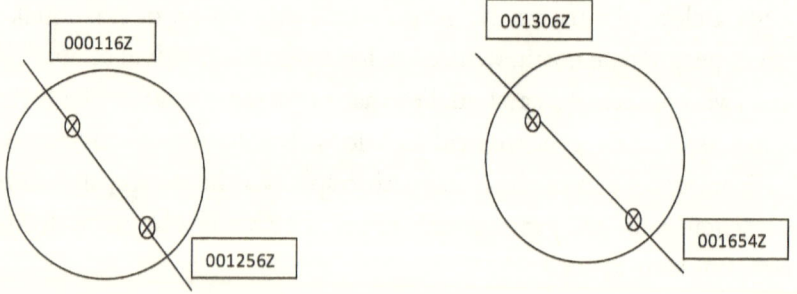

At 000116Z, a magnitude 1.6 tremor struck 44 kilometres south-south-west of Cordova, Alaska. At 001256Z a second registered quake for the day, measured at 2.3, struck the British Virgin Islands, 7,750 kilometres away. The orientation of these two epicentres is about 328.36 degrees. The foci of the quake started in the north and then shifted southward.

We can assume that this straight line is active for about eleven minutes because by 001306Z, the epicentre of the quake shifted back to the north with a quake near Old Iliamna in Alaska. This third quake occurred precisely on the same straight line.

By 001654Z, a new tremor was recorded at the Northern Mid-Atlantic Ridge. We see that the quake moved back to the south. This new line had an orientation of 330.62 degrees. It shifted 2.26 degrees away from the earlier alignment. This line was active for about four minutes.

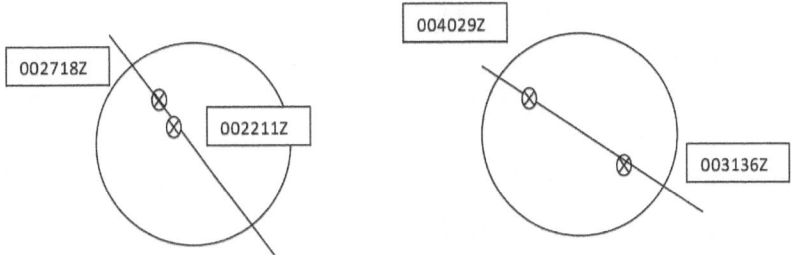

After a period of about five minutes, a fresh tremor hit 65 kilometres south of Cantwell, Alaska at 002211Z. By 002718Z, another quake strike farther north with the epicentre 96 kilometres north-north-west of Talkeetna, Alaska. The orientation of these two new epicentres would place the line in a 287.57-degree orientation to true north. The distance between the two epicentres is only 110 kilometres.

After a gap of about nine minutes, another quake was registered southward at 003136Z, at the Northern Mid-Atlantic Ridge. At 004029Z, a low-magnitude tremor struck near Fishhook, Alaska. The quakes appeared to be alternating their locations north and south from each other.

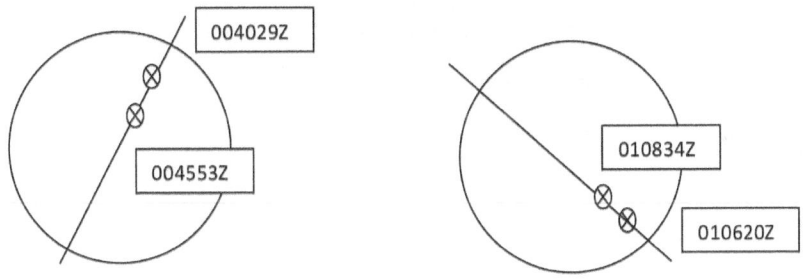

After about nine quiet minutes with no tremor recorded, at 004029Z, a magnitude 1.0 quake was detected 20 kilometres north of Fishhook,

Alaska. This was followed at 004553Z in the south-west by another low-magnitude quake, 76 kilometres south of King Salmon, Alaska. These two epicentres were separated by a distance of 591 kilometres and were aligned at 39.51 degrees to true north.

This new alignment, when compared to the previous one, has changed or flipped to its opposite side of true north. This new alignment would completely rearrange the next epicentre on a new orientation.

Unfortunately, there was a gap of twenty minutes with no earthquakes recorded. It could be possible that one or two low-magnitude tremors occurred in some other regions (perhaps in the Far East) during this time period, but it was not reported or documented in the USGS Earthquake Catalog.

After the twenty-minute period, two quakes struck South America, at 010620Z near Valparaiso and at 010834Z near Cartagena. The distance was only 36.45 kilometres. We shall call this type of event a close-proximity double quake. When predicting earthquakes, we must be aware of these type of quakes because they do not conform to the expected far-apart quakes. The gap between these two quakes was about two minutes. The orientation of the tremors was 304.52 degrees.

This new orientation would place the quakes back in the general north-west–south-east orientation. We will see later on that the north-west–south-east orientation is more dominant than the north-east–south-west orientation. This should be kept in mind when predicting earthquakes for the preliminary earthquake prediction protocol mentioned at the end of this handbook.

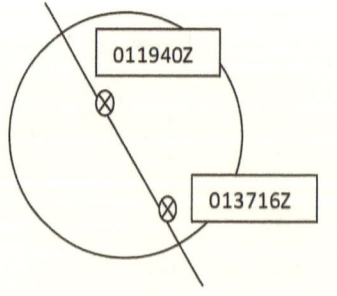
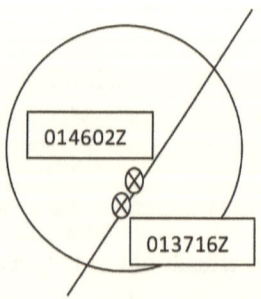

About thirteen minutes after the Chilean earthquake, a slight tremor hit 2 kilometres south of Hillsborough, California, at 011940Z. The line is still in the north-west–south-east orientation. Eighteen minutes later, another quake was recorded back south, in Chile at 013716Z near Valparaiso. Take note that the quake is always jumping back and forth from north to south, and vice versa.

At 014602Z, another close-proximity double quake struck north-east of the earlier quake. The distance between these two quakes was only 9.48 kilometres. The compass orientation, however, flipped to the opposite side at 83.84 degrees to true north.

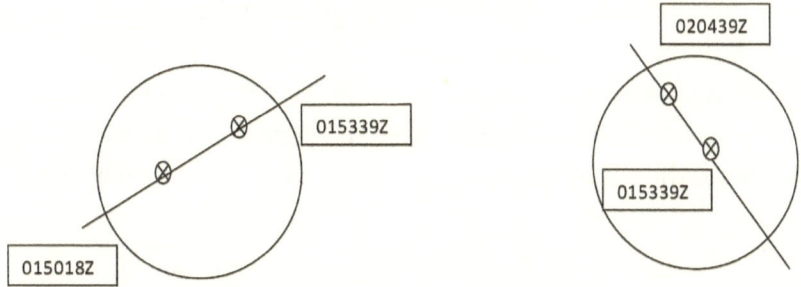

Four minutes after the 014602Z tremor, the orientation was still in the general north-east–south-west alignment with a minor tremor registered near Volcano, Hawaii, at 015018Z. Three minutes after the Hawaiian tremor, a low-magnitude quake was felt at 015339Z, 21 kilometres north-east of Soledad, California. The line was still holding in the north-east–south-west orientation.

Eleven minutes later, a tremor was recorded at 020439Z, 15 kilometres east of Willow, Alaska. The line, as we can see, flipped back to the north-west–south-east orientation.

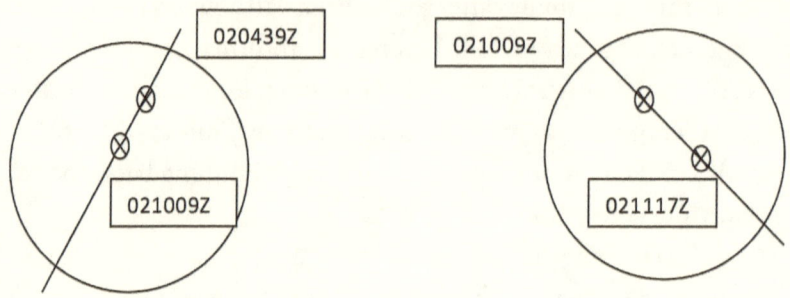

Six minutes after the 020439Z tremor, another quake was felt, again in Alaska but to the south-west at an epicentre located 237 kilometres south-east of Akutan at 021009Z. The general orientation of the line was still in the north-east–south-west orientation.

Just one minute later, another quake was registered at 021117Z, 29 kilometres east-north-east of West Yellowstone, Montana. The line orientation flipped in the opposite direction in a short span of time.

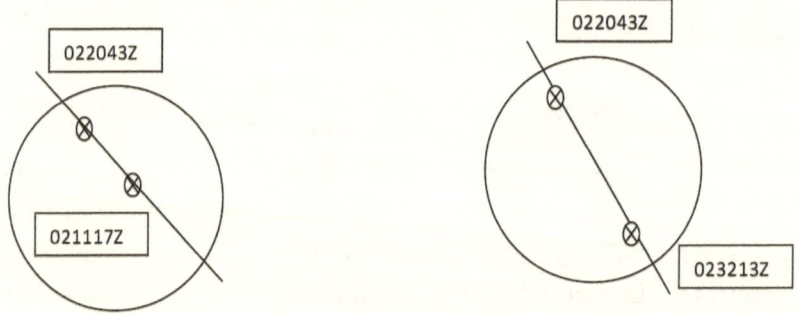

After the tremor in Montana, about nine minutes later, the quake struck back north at 022043Z, 90 kilometres west-north-west of Talkeetna, again in Alaska. The line orientation was still in the north-west–south-east direction.

Twelve minutes later, a 4.1 magnitude earthquake struck far south in Argentina at 023213Z. The quake moved south, but the line orientation still held in the general north-west–south-east direction.

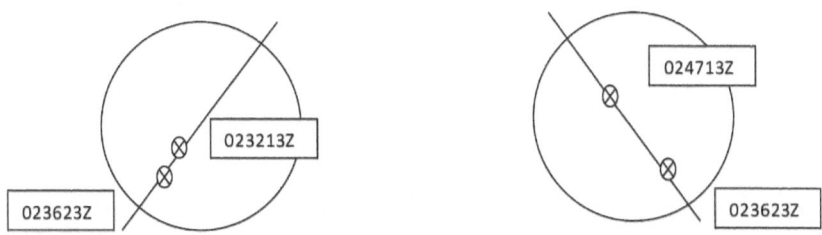

Just four minutes after the Argentinean quake, another tremor was felt to the south-west in Chile at 023623Z. The orientation of the line flipped to the opposite side. The distance between the two epicentres was 1,098 kilometres.

Eleven minutes later, a quake struck north, 13 kilometres south-east of Chester, California, at 024713Z. The orientation was back to the north-west–south-east direction.

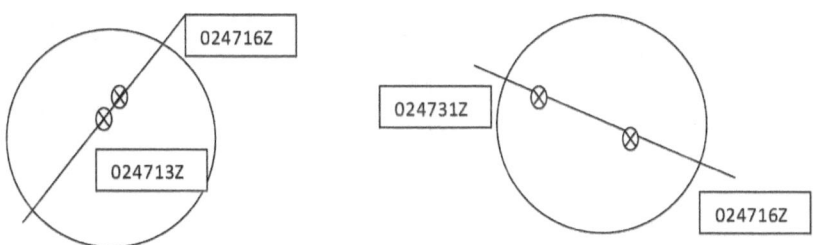

Three seconds after the Chester tremor, another quake struck Oregon at 024716Z. Then fifteen seconds later, the quake moved north-west to strike Alaska at 024731Z. The line flipped back to the north-west–south-east direction.

We can see that during a time frame of two hours forty-seven minutes, 67 percent of the time the orientations was in the general north-west–south-east direction, whereas only 33 percent was in the north-east–south-west orientation. We can use this figure as a guideline of which orientation to expect for the next quake.

To get a clearer picture of what actually transpired, please refer to Table 2 below. It lists 100 global epicentres registered on 29 April

2017. It is estimated that more than 30 percent of tremors between magnitudes 1.0 to 2.4 for areas outside of the United States are not listed in the USGS Earthquake Catalog.

Holding time in the Table refers to the time between two successive earthquakes. It is assumed that this time frame is the time whereby the energy beam swept across the corridor several times for a certain period before changing to a new orientation.

The orientation is taken in relation to the true north. We can see two opposing alignments (i.e., one on the left of the true north and one on the right), as depicted in Figure 10 below. The change of the line alignment from the north-west to the north-east or vice versa is considered a flip. Changes in the line alignments but still holding in the north-west or north-east is called an angle shift.

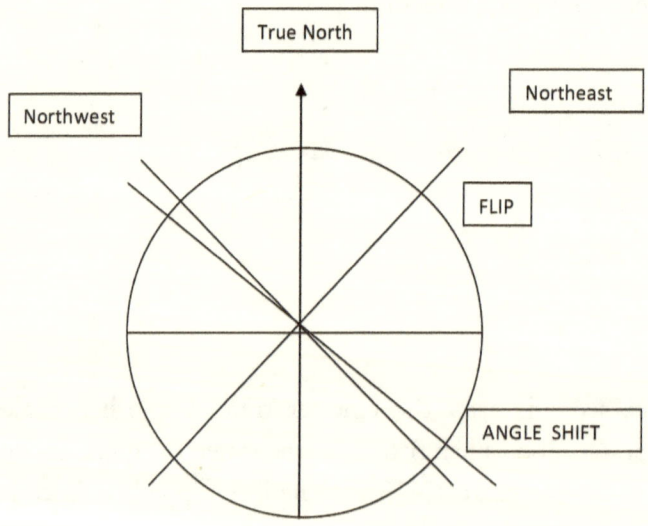

*Figure 10. Orientations of SLAs and its flip angles.*

Table 2: Spatial-temporal distribution of earthquakes in time and space for 29 April 2017.

| Time A | Region A | Time B | Region B | Distance (km) | Orientation (Degree) | Holding Time (Min) |
|---|---|---|---|---|---|---|
| 000116Z | Cordova, Alaska | 001256Z | Br. Virgin Is | 7750 | 328.36 | 11 |
| 001306Z | Old Iliamna, Alaska | 001654Z | N. Mid-Atlantic | 8536 | 330.62 | 3 |
| 002211 | Cantwell, Alaska | 002718Z | Talkeetna, Alaska | 110 | 287.57 | 5 |
| 004029Z | Fishhook, Alaska | 004553Z | King Salmon, Alaska | 591 | 39.51 | 5 |
| 010620Z | Valparaiso, Chile | 010834Z | Cartagena, Chile | 36.45 | 304.52 | 2 |
| 011940Z | Hillsborough, California | 013716Z | Valparaiso, Chile | 9411 | 322.21 | 18 |
| 014602Z | Valparaiso, Chile | 015018Z | Volcano, Hawaii | 10577 | 289.83 | 4 |
| 015339Z | Soledad, California | 020439Z | Willow, Alaska | 3431 | 333.81 | 11 |
| 021009Z | Akutan, Alaska | 021117Z | W. Yellowstone, Montana | 3907 | 301.85 | 1 |

| | | | | |
|---|---|---|---|---|
| 021430Z | Talkeetna, Alaska | 247 | 344.02 | 6 |
| 023213Z | Argentina | 1098 | 25.07 | 4 |
| 024713Z | Chester, California | 222 | 34.08 | 3 sec |
| 024731Z | Sitkin Is., Alaska | 4708 | 307.19 | 2 |
| 031022Z | Puerto Rico | 7921 | 329.23 | 10 |
| 032408Z | Sinabang, Indonesia | 14329 | 37.79 | 12 |
| 034524Z | N. Mid-Atlantic | 5129 | 297.89 | 3 |
| 035703Z | Harper, Kansas | 1801 | 79.55 | 3 |
| 040446Z | Chitina, Alaska | 789 | 48.40 | 16 |
| 042321Z | Haines Junc., Canada | 2920 | 337.90 | 4 |
| 043101Z | Anchor Point, Alaska | 8488 | 31.42 | 2 |
| 043740Z | Homer, Alaska | 134 | 343.59 | 2 |
| 044516Z | Tiptonville, Tennessee | 7925 | 345.27 | 5 |

| | | | | |
|---|---|---|---|---|
| 045539Z | Cabazon, California | | 5376 | 299.34 | 5 |
| 054047Z | Redoubt Vol., Alaska | 050034Z Virgin Islands | 163 | 346.24 | 6 |
| 055757Z | Ester, Alaska | 054600Z Homer, Alaska | 395 | 4.48 | 8 |
| 060805Z | Saumlaki, Indonesia | 060508Z Meadow Lakes, Alaska | 9772 | 30.69 | 6 |
| 061701Z | Afghanistan | 061425Z Larsen Bay, Alaska | 11244 | 359.57 | 42 sec |
| 062101Z | Old Fateful Geyser, Wyoming | 061743Z Old Fateful Geyser, Wyoming | 0.94 | 352.85 | 41 sec |
| 063035Z | Anchorage, Alaska | 062142Z Old Fateful Geyser, Wyoming | 218 | 271.59 | 10 |
| 064407Z | Old Iliamna, Alaska | 064042Z Valdez, Alaska | 707 | 223.01 | 5 |
| 065525Z | Fair Haven, Vermont | 064924Z Sand Point, Alaska | 5208 | 322.40 | 23 |
| 072451Z | Unalaska, Alaska | 071837Z Talkeetna, Alaska | 1177 | 233.51 | 1 |
| | | 072532Z Redoubt Vol., Alaska | | | |

| | | | | | | |
|---|---|---|---|---|---|---|
| 074524Z | Fayzabad, Afghanistan | 074822Z | Willow, Alaska | 8383 | 326.81 | 3 |
| 080134Z | Tonga | 081134Z | The Geysers, California | 8080 | 38.91 | 10 |
| 081247Z | Lakeport, California | 081534Z | R. P. Verdes, California | 678 | 332.27 | 3 |
| 081832Z | Amatignak Is., Alaska | 082300Z | The Geysers, California | 4491 | 307.16 | 5 |
| 082630Z | Anchor Point, Alaska | 083042Z | Valparaiso, Chile | 12623 | 326.95 | 4 |
| 083625Z | Itoman, Japan | 085013Z | Pangai, Tonga | 8074 | 305.68 | 14 |
| 085401Z | Valparaiso, Chile | 090213Z | Volcano, Hawaii | 10559 | 289.93 | 8 |
| 090456Z | Redoubt Vol., Alaska | 091924Z | Br. Virgin Is. | 8079 | 328.75 | 15 |
| 094815Z | Medford, Oklahoma | 094942Z | Beatty, Nevada | 1658 | 276.82 | 1 |
| 095858Z | Medford, Oklahoma | 095918Z | Mammoth Lakes, California | 1881 | 278.98 | 20 sec |
| 100125Z | Polson, Montana | 100212Z | Tambakrejo, Indonesia | 13986 | 36.45 | 1 |

| | | | | |
|---|---|---|---|---|
| 100633Z | La Cumbre, Colombia | 100858Z | Dawson, Canada | 8371 | 337.14 | 2 |
| 101302Z | Canwell, Alaska | 102327Z | Anchor Point, Alaska | 370 | 11.97 | 10 |
| 102632Z | The Geysers, California | 103116Z | Nikiski, Alaska | 3134 | 330.40 | 5 |
| 104021Z | Dominican Rep | 104945Z | Cantwell, Alaska | 7705 | 331.52 | 9 |
| 105321Z | DR Congo | 105603Z | North Atlantic | 7060 | 325.21 | 3 |
| 111157Z | Hawthorne, Nevada | 111233Z | Br. Virgin Is. | 5680 | 304.82 | 1 |
| 111548Z | Talkeetna, Alaska | 112251Z | Pahala, Hawaii | 4886 | 2.76 | 7 |
| 112438Z | Talkeetna, Alaska | 112607Z | Big Lake, Alaska | 182 | 341.04 | 2 |
| 114244Z | Talkeetna, Alaska | 114250Z | Unalaska, Alaska | 1331 | 36.77 | 6 sec |
| 114750Z | Talkeetna, Alaska | 115551Z | Pahala, Hawaii | 4929 | 2.59 | 8 |
| 120609Z | Anchor Point, Alaska | 121054Z | False Pass, Alaska | 894 | 39.16 | 4 |
| 121836Z | US Virgin Is. | 122402Z | Willow, Alaska | 7922 | 330.24 | 6 |

| | | | | | | |
|---|---|---|---|---|---|---|
| 122756Z | Chignik Lake, Alaska | 123234Z | Nishinoomote, Japan | 6039 | 40.38 | 5 |
| 124119Z | Minatitian, Mexico | 130108Z | Talkeetna, Alaska | 6022 | 335.89 | 20 |
| 130445Z | Old Iliamna, Alaska | 131028Z | Granite Falls, Washington | 2408 | 314.28 | 6 |
| 131028Z | Granite Falls, Washington | 131630Z | Burgos, Philippines | 10816 | 38.76 | 6 |
| 132149Z | Anchorage, Alaska | 133546Z | Solomon Islands | 8971 | 21.38 | 14 |
| 134034Z | Nishinoomote, Japan | 134118Z | Pacifica, California | 9244 | 303.78 | 1 |
| 134153Z | Baetov, Kyrgyzstan | 135155Z | Banning, California | 11667 | 350.96 | 10 |
| 135945Z | US Virgin Is. | 140330Z | Larsen Bay, Alaska | 8427 | 326.03 | 4 |
| 142411Z | Nevada | 142528Z | Talkeetna, Alaska | 3335 | 328.46 | 1 |
| 143004Z | Tambakrejo, Indonesia | 143613Z | S. Fiji Islands | 7332 | 270.86 | 6 |
| 143806Z | Redoubt Vol., Alaska | 144643Z | Whittier, Alaska | 322 | 75.18 | 8 |
| 145614Z | Waimea, Hawaii | 150553Z | Papua New Guinea | 6221 | 62.90 | 9 |

| | | | | | |
|---|---|---|---|---|---|
| 151424Z | Puerto Rico | 151900Z | The Geysers, California | 5957 | 305.47 | 5 |
| 152301Z | US Virgin Is. | 153051Z | Cempa, Indonesia | 18676 | 22.73 | 7 |
| 153937Z | King Salmon, Alaska | 154554Z | Tambakrejo, Indonesia | 11020 | 32.89 | 6 |
| 155901Z | Jorok Dalam, Indonesia | 155921Z | Redoubt Vol., Alaska | 10720 | 30.38 | 20 sec |
| 160350Z | Anchor Point, Alaska | 160814Z | Junction, Utah | 3707 | 323.50 | 5 |
| 161823Z | Avalon, California | 163425Z | Whittier, Alaska | 3731 | 334.56 | 16 |
| 163941Z | Sterling, Alaska | 163944Z | Ilave, Peru | 11068 | 330.22 | 3 sec |
| 165028Z | Tobelo, Indonesia | 165046Z | Nikiski, Alaska | 9135 | 28.48 | 18 sec |
| 165058Z | Hengchun, Taiwan | 165943Z | Talkeetna, Alaska | 7814 | 28.74 | 9 |
| 170700Z | Delta, Mexico | 172846Z | Talkeetna, Alaska | 4158 | 334.50 | 21 |
| 173733Z | Ngulung Wetan, Indonesia | 174741Z | Balanggonan, Philippines | 2231 | 41.70 | 10 |
| 175128Z | Nephi, Utah | 175221Z | Padangsidempuan, Indonesia | 14530 | 32.12 | 1 |

| | | | | |
|---|---|---|---|---|
| 175843Z | Masachapa, Nicaragua | 181920Z | Running Springs, California | 3989 | 314.77 | 21 |
| 182807Z | Talkeetna, Alaska | 183518Z | Ridgemark, California | 3606 | 334.77 | 7 |
| 183923Z | Bitung, Indonesia | 185206Z | Mammoth Lakes, California | 12156 | 48.81 | 13 |
| 185515Z | Mammoth Lakes, California | 190454Z | Gustavus, Alaska | 2703 | 335.51 | 9 |
| 190846Z | Anacortes, Washington | 191228Z | Point Hope, Alaska | 1740 | 317.50 | 4 |
| 191818Z | Searles Valley, California | 191841Z | Milford, Utah | 493 | 54.71 | 23 sec |
| 192859Z | Hawthorne, Nevada | 194049Z | Nishinoomote Japan | 9464 | 305.76 | 12 |
| 194150Z | Nishinoomote, Japan | 194204Z | Balangonan, Philippines | 2888 | 12.61 | 1 |
| 200501Z | Bobong, Indonesia | 201536Z | Old Iliamna, Alaska | 9712 | 30.93 | 10 |
| 201809Z | The Geysers, California | 201817Z | The Geysers, California | 0.16 | 44.02 | 8 sec |

| | | | | |
|---|---|---|---|---|
| 202259Z | Nishinoomote, Japan | 204359Z | Nevada | 9452 | 307.12 | 21 |
| 205830Z | Lluta, Peru | 210358Z | Enterprise, Utah | 7335 | 324.72 | 5 |
| 210618Z | Br. Virgin Is. | 211656Z | Anchor Point, Alaska | 8139 | 328.27 | 10 |
| 212212Z | Hollister, California | 212619Z | Big Lake, Alaska | 3394 | 333.37 | 4 |
| 213721Z | Ferndale, California | 221556Z | Ilapel, Chile | 9604 | 322.69 | 38 |
| 221704Z | Saumlaki, Indonesia | 222039Z | Pahala, Hawaii | 8650 | 69.23 | 3 |
| 222552Z | Anthony, Kansas | 222830Z | Nsunga, Tanzania | 13488 | 313.97 | 3 |
| 223333Z | Solomon Islands | 223736Z | Cantwell, Alaska | 9111 | 19.82 | 4 |
| 224103Z | Pine Valley, California | 224119Z | Ishigaki, Japan | 11042 | 306.97 | 16 sec |
| 224141Z | Ishigaki, Japan | 225930Z | Sulangan, Philippines | 1466 | 349.10 | 18 |
| 231522Z | Vanuatu | 234006Z | Talkeetna, Alaska | 9233 | 17.99 | 25 |
| 234917Z | Mariana Islands | | | | | |

From the above table, based on the holding time frames, we can classify the earthquake frequency into three categories: Fast for less than one minute, Medium for two to ten minutes, and Slow for greater than eleven minutes. We summed up the observed times as shown in Table 3 below.

*Table 3: Observed time frame for the frequency of earthquakes.*

| Timing of Successive Quakes | Fast<br>Less than 1 min. | Medium<br>2 to 10 min. | Slow<br>More than 11 min. |
|---|---|---|---|
| Holding Time for Each Corridor | 13.5% | 66.5% | 20% |
| Mean | 37.2 sec/quake | 5.4 min/quake | 17.2 min/quake |

We assumed from the above figures that the rotational rate of Source X on a normal day is most probably in the range of two to ten minutes. From this we can expect that earthquakes will occur one after the other between a two- to ten-minute time frame with the average of one earthquake every five minutes.

Occasionally it could either speed up or slow down. The possibility of this happening is about 10–20 percent of the time. On an "abnormal" day when there is a major quake followed by numerous aftershocks, the number of quakes per minute will increase drastically, but for a limited time frame only before it will settle back to the average of one tremor every five minutes.

What is notable from these observations is that the straight line alignments will hold their course only for a certain period of time, usually between two to ten minutes, before they change their compass orientation. Although their angle of orientation changes, their basic pattern was still the same—they will flip either to the left or right.

This means that after a certain period, an alignment will flip to the opposite side. We will call this the Seesaw Effect because the line changes its orientation back and forth like a seesaw. This important effect can help us to determine the next earthquake on the new orientation straight line.

Step 8

# THE SEESAW EFFECT OF EARTHQUAKE EPICENTRE DISTRIBUTION

We understand how a seesaw works, and it takes two to play. As one party moved upward, the other side moved down. In other words, the plank always moved in the opposite side when one party exerts extra pressure. This same effect can be seen in the pattern of the global earthquake distribution in time and space.

In Figure 11a, we see the distance between two points of a seesaw, *a1* and *a2,* are the same as point *b1* and *b2*. This is because the pivot is at the centre of the plank.

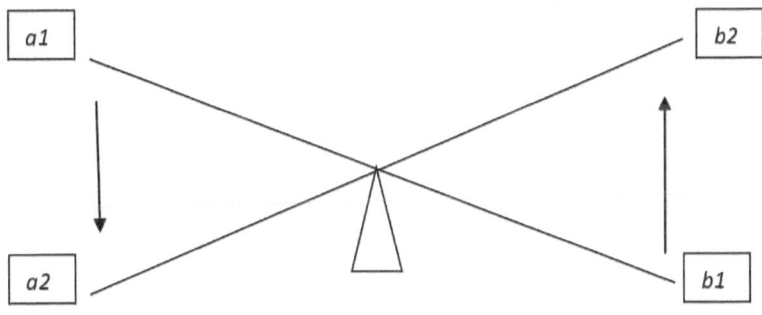

*Figure 11a: Seesaw effect with the pivot in the centre.*

If the pivot is slightly off the centre, the distance between *a1* and *a2* becomes much shorter or nearer. The distance between *b1* and *b2*, however, will be farther. Refer to Figure 11b below.

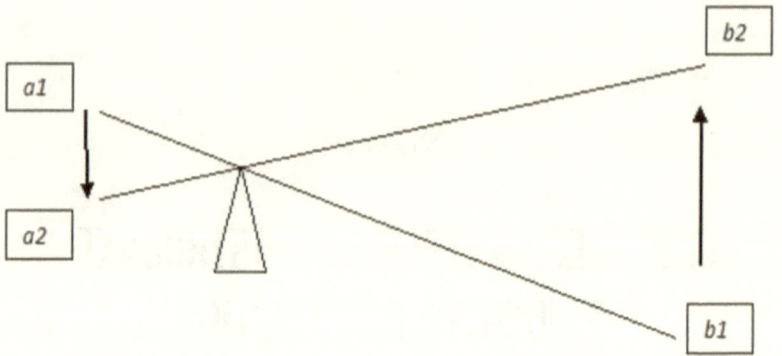

*Figure 11b: Seesaw effect with the pivot off the centre.*

This same effect can be seen for the distribution of earthquake epicentres. The interchanging orientation lines of earthquake epicentres usually will not cross at the centre. This will result with earthquakes being concentrated or clustering at the *a1* and *a2* locations. Because the locations of *a1* and *a2* could be quite close, we could miss seeing that there are actually two separate clusters of earthquake epicentres as both clusters intertwine into each other.

On 29 April 2017, for example, there were several geographical clusters of earthquake epicentres. If we take the Northern Mid-Atlantic Ridge as point *b2* and the British Virgin Islands as point *b1*, then its reciprocal effect is in Alaska with point *a1* in the Talkeetna region and point *a2* in the Cordova/Old Iliamna region. Its geographical seesaw orientation is shown in Figure 12 below from Google Earth Pro.

Any shifts in the alignment angles would include a Puerto Rican tremor reciprocating with the other Alaskan quakes. Due to this seesaw effect, in one day, even without a major quake and its so-called aftershocks, Alaska had many tremors on its own.

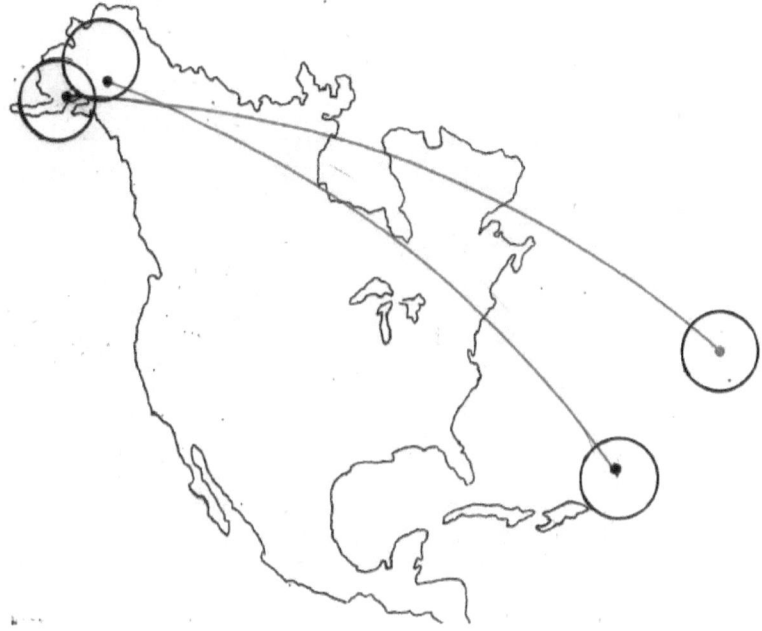

*Figure 12: The seesaw effect for earthquake clusters in Alaska, Mid-Atlantic Ridge, and Virgin Islands Regions.*

This same seesaw effect can be seen in the other regions as well, where there are fault lines present. The other earthquake clusters on 29 April 2017, are in the Valparaiso, Chile, region. Its reciprocal region is also in Alaska. Some epicentres clusters in California also ended in Alaska. For this reason, most of the quakes on that date were found clustering in Alaska.

Other clusters of tremors were in the Nishinoomote region of Japan and the area south-west of Java, Indonesia. Tremor clusters in Hawaii further added to the reciprocal number of quakes in Alaska.

The question now is, "Why do we see this seesaw effect in the distribution of earthquakes epicentres?" The answer lies with the analogy below. It has to do with the behaviour of the contributing source or Source X. It appears that it is behaving like a lighthouse beam on a gyroscope.

A lighthouse beam or light will execute a 360-degree rotation. When seen from the side, it described a straight line horizontal to the land. Now, if we place the light on a rotating gyroscope, the light will still show a horizontal line when the gyroscope is spinning rapidly and its position stabilises at a horizontal position. Once the gyroscope loses its speed, it will dip down, and the light will beam at an angle to the horizontal (refer to Figures 13a and 13b).

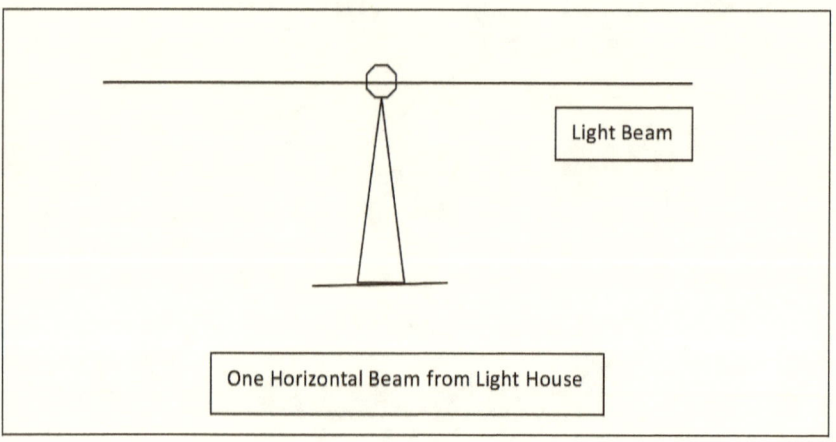

*Figure 13a: Horizontal line for a steady base light beam.*

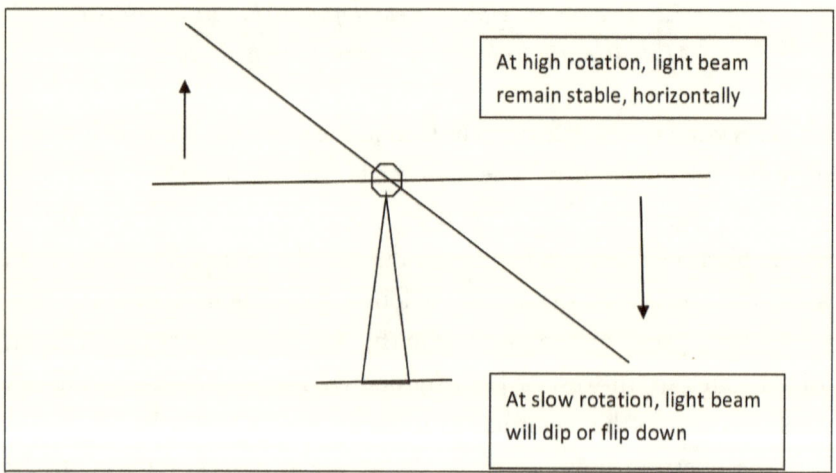

*Figure 13b: Horizontal and reclining light beams on a gyroscope.*

The changing straight lines of earthquake orientation seem to mimic the behaviour of the light beam on a gyroscope. The presence of an astrophysical powerhouse in the sky could exhibit the same action due to its changing speed of rotation. This speed is assumed to be dictated by its energy emission, which is giving it the momentum to spin either slow or fast. This in turn will determine its dip or the angle of the beam sweep. We shall refer this as its flip angle.

The energy beam from Source X that crosses the face of the Earth will flip its orientation angle from time to time due to its speed variation.

If the theory is correct, we can expect to see earthquakes adhering to a distribution pattern like this:

1. Two, three, or more earthquakes will be aligned in a straight line corridor. Once in a while, it is possible to see three or more epicentres on a perfect straight line.
2. The next impending quake will strike several degrees away from this straight line corridor, if the angle shift is small.
3. If the angle of the straight line alignment, for example, is in the north-west–south-east orientation, it could flip to the opposite side when there is a change in its rotational speed. Earthquakes then will be in a new alignment (i.e., in the general north-east–south-west orientation).
4. The times of earthquake occurrences will be directly related to the speed of rotation of Source X.

To make an accurate earthquake prediction, we need to determine the rotational speed of Source X, its tilt angle, and the length of two reciprocal earthquakes. At present we can make only a mean estimate of its changing speed modes and see how much the tilt angle is.

Step 9

# DETERMINE THE SHIFT AND FLIP ANGLE OF SOURCE X

Earthquakes seem to occur in pairs—not in the sense that two quakes strike at the same time but later and on the same alignment. The alignments of two pairs of earthquake epicentres opposite each other on a north-east or north-west orientation will give us the flip angle.

The flip angle refers to its compass orientation with respect to true north. To justify a flip, one line orientation must be left of true north, and the other must be on its right. If there is a minor angle shift between two orientation lines on the same side of true north, then we do not called it a flip. It will be called an angle shift.

The holding period was taken as the average time for the north-west–south-east SLAs and the north-east–south-west SLAs to hold its course on the said alignment before flipping to its opposite side.

The breakdown for the SLA orientations and its characteristics is given in Table 4 below.

*Table 4: SLA orientations with angle shifts, flip angles, and holding times for 29 April 2017.*

| SLA Characteristics | NW-SE Orientation | NE-SW Orientation |
|---|---|---|
| No. of SLA (Straight Line Alignment) | 64.6% | 34.4% |
| Mean Orientation Line | 317.67 deg | 35.72 deg |
| Mean Flip Angle NW to NE | — | 77.34 deg |
| Mean Flip Angle NE to NW | 71.64 deg | — |
| Mean Angle Shift | 41.5 deg | 17.64 deg |
| Mean Holding Time | 20.26 min | 9.65 min |

There were a total of 99 SLAs for 29 April 2017, and 64 (or 64.6 percent) were oriented in the north-west–south-east alignment. The remaining 34 SLAs (34.4 percent) were in the north-east–south-west alignment. The north-west–south-east alignment appears to be more dominant.

The mean flip angle from north-west–south-east SLAs to north-east–south-west alignments is about 77 degrees. The opposite flip from the north-east–south-west alignments is about 72 degrees. Overall, the mean flip angles appear to be holding at 72–77 degrees.

The mean holding time is longer in the north-west–south-east orientation, which is 20 minutes compared to the north-east–south-west alignment, which is just about half the time at 9.65 minutes.

From the table above, we can use 317 degrees as the mean reference SLA to predict earthquakes on the north-west–south-east line. For the opposite north-east–south-west SLA, use 36 degrees as the line to determine where the next tremor will occur.

Figure 14 below gives a graphic illustration how the orientation lines were formed. If we assume that Source X emits a laser pointer–like beam that describes a straight line as it sweeps across the face of the globe, a straight line will start at point A and will terminate at point B as the beam recedes away into space. Earthquakes can occur at either point A or B.

When this beam completed its circular rotation, the beam will again be pointed towards Earth. If the source is rotating steadily like a fast-spinning top, its line sweep will cross points A to B again. But this is unlikely to happen on the second pass because there could be some minor change in its speed of rotation and a slight change in its tilt; the beam will be slightly offset from the earlier straight line. For this reason, we can see earthquake epicentres clustering in the SLCs.

It will not pass directly along the straight line of the first pass. In other words, it is rare to get three or four epicentres on a perfect straight line. However, all the tremors would still fall into 5-kilometre-confined straight line corridors (SLCs)

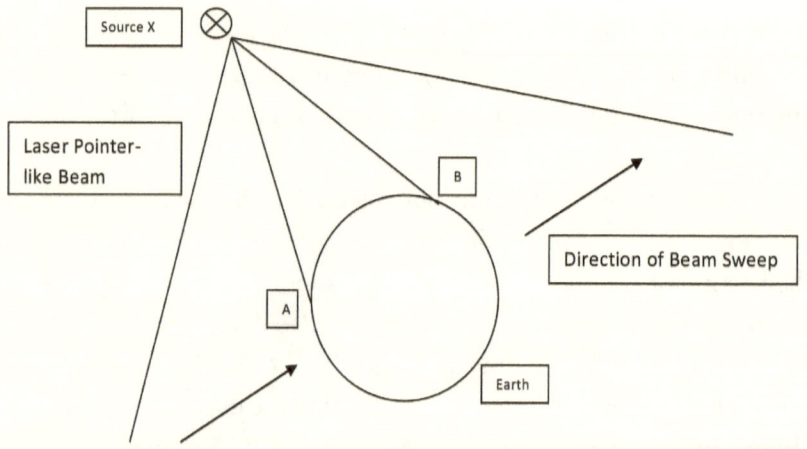

*Figure 14: Laser pointer–like beam sweep across the face of the Earth.*

If the energy beam dips or flip at a considerable angle, we can see a new orientation line opposite to the earlier one. When this happens, we can see four minor clusters of earthquakes at the ends of both orientation lines. As depicted in Figure 15 below, earthquakes tend to cluster in a small geographical areas at A, B, C, and D.

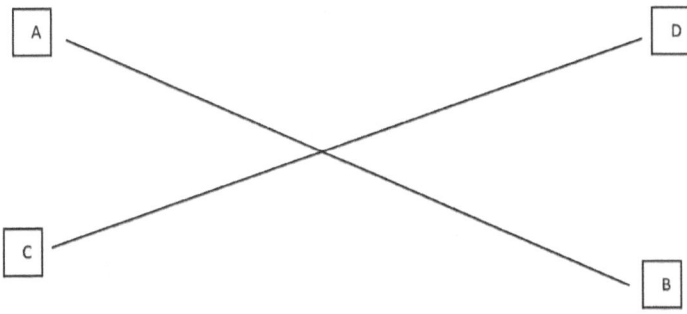

*Figure 15: Earthquake clusters at the four ends of the SLAs.*

Based on these observations, we can say the following:
1. The orientation lines do exist and are part of the earthquake distribution pattern.
2. The clustering of earthquakes at the ends of the two lines (SLCs) do occur and is another important characteristic of earthquake distributions.
3. The orientation line changes its compass orientation with time, and hence this changes the geographical locations of the earthquake epicentres. The flip angle is important to determine where the next quake will occur.
4. The location of the earthquake epicentres seems to alternate each other. For example, if the quake struck at A in the north, the next one will be at B in the south. After that, it will changed back to the north, and so on.
5. As the orientation line changes its angle, the next quake will strike at point C northward, followed by point D later, as it tends to shift the focus back southward (refer to Figure 16 below). This will happen in five to seventeen minutes, or a lesser time frame of less than 60 seconds. For cases where we see quakes occurring more than ten to twenty minutes, an error could arise because there could possibly be other undocumented low-magnitude tremors not listed in the USGS Catalog.

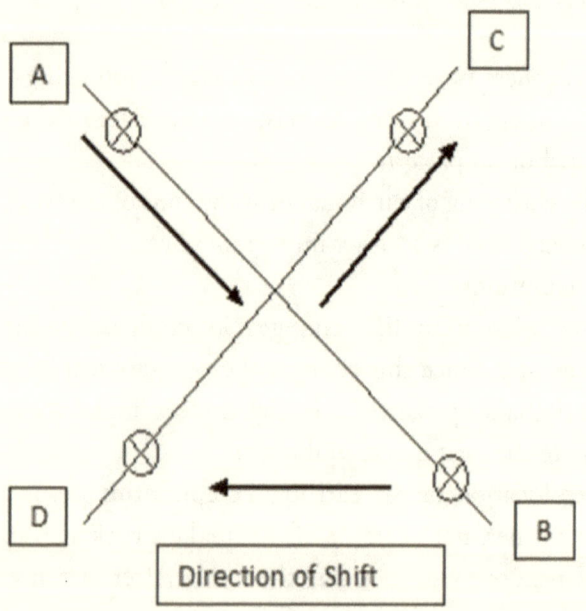

*Figure 16: Shifting earthquake epicentres due to the seesaw effect.*

Step 10

# SEE HOW AND WHY LARGE-MAGNITUDE EARTHQUAKES HAPPEN

Large or mega magnitude earthquakes are the primary cause for the large loss of lives and the damages inflicted to infrastructures and properties. These types of quakes are sometimes accompanied with a tsunami if the epicentre is located in the seabed.

The 26 December 2004, great Sumatran earthquake measured 9.1, had its epicentre off the coast, and took its toll with the devastating tsunami that followed. This resulted in over 230,000 deaths.

Seismologists attributed large earthquakes to one tectonic plate being forced underneath another plate. When this happen, it is considered the most powerful earthquake on Earth.

Something of a sudden nature and force must have incited the plates to move against the grip of gravity and mantle cohesion. The plates did not moved on their own.

To forecast any major quake in the future, the foundation laid in this book could be of some value. We have shown that earthquake epicentres are arranged or aligned in a straight line corridor (SLC) about 5 kilometres wide. This means that most quakes tend to occur in the straight line corridor, especially at both its ends. Any impending quake can be expected to occur in any corridors because there tends to be several of them in a twenty-four-hour period before a major quake strike.

Let us take a look for a twenty-four-hour period before the quake on 26 December 2004, with its epicentre located off the north coast of western Sumatra, Indonesia. The epicentre is identified as 005853Z (its time of occurrence and not its geographical coordinates).

Table 5 below lists some of the SLCs that have three or more epicentres along the line from the United States to Sumatra, Indonesia.

Table 5. *The eight identified SLCs convergent to the Sumatra epicentre.*

| Corridor Time/Location | Epicentre in Corridor | Dev. (km) | Epicentre in Corridor | Dev. (km) | Epicentre in Corridor | Dev. (km) | Length of Corridor (km) | Orientation (Deg) |
|---|---|---|---|---|---|---|---|---|
| 001523Z (26 Dec) – 005853Z | 220949Z/ Nevada | 0.01 | 013843Z/ Central California | 2.66 | 000309/ Central California | 4.65 | 14,217 | 314.26 |
| 002501Z (26 Dec) – 005853Z | 002501Z Central California | | 113448Z/ Central California | 1.88 | 005853Z Northern Sumatra | 0.00 | 14,205 | 314.13 |
| 164444Z (Mexico) – 005853Z | 132916Z Central California | 1.49 | 010136Z Central California | 1.20 | 181432Z Central California | 1.16 | 14,758 | 312.81 |
| | 014437Z Central California | 4.09 | 132011Z Central California | 1.62 | 060437Z Central California | 2.01 | | |
| | 014834Z Central California | 1.06 | 000640Z Central California | 1.13 | 222038Z Central California | 3.41 | | |
| 234405Z (25 Dec) – 005853Z (26 Dec) | 163719Z Northern California | 5.08 | 094950Z Northern California | 4.45 | 191706Z Northern California | 1.71 | 14,029 | 311.21 |

| Time Range | Event 1 | Val | Event 2 | Val | Event 3 | Val | | |
|---|---|---|---|---|---|---|---|---|
| 192633Z (25 Dec) – 005853Z (26 Dec) | 231235Z Idyllwild, California | 2.42 | 134931Z Fontana, California | 1.43 | 045336Z Fontana, California | 3.29 | 14,701 | 313.67 |
| | 045413Z Buttonwillow, California | 2.93 | | | | | | |
| 182300Z (25 Dec) – 005853Z (26 Dec) | 075720Z Central California | 0.00 | 000309Z Central California | 0.27 | 013843Z Central California | 2.21 | 14,219 | 314.12 |
| | 042755Z Central California | 0.80 | 113448Z Central California | 3.03 | | | | |
| 001933Z (25 Dec) – 005853Z (26 Dec) | 205820Z Northern California | 0.15 | 162501Z Northern California | 0.56 | 004749Z Northern California | 1.16 | 13,888 | 310.43 |
| | 001018Z Northern California | 0.57 | 101107Z Northern California | 0.62 | 221353Z Northern California | 0.00 | | |
| 235446Z (24 Dec) Sumatra – 203809Z (24 Dec) | 235446Z Java, Indonesia | 0.00 | 005853Z Sumatra | 4.27 | 203809Z Sumatra | 0.00 | 2,070 | 306.43 |

As can be seen in the table above, seven of the identified SLCs originated from the United States and Mexico. They converged on a bearing that passed over the epicentre of the Sumatran earthquake. This is like aiming seven guns at a single target. Imagine the impact. And with respect to the impending quake in Indonesia, it came with a bang, culminating in a magnitude 9.1 earthquake.

As a result of the megaquake, the violent crustal displacements that took place caused a surge in the oceanic waves. The tsunami washed everything in its wake and did more damage than the quake itself.

There were six clusters of earth tremors in a twenty-four-hour period in the United States for magnitude 1.0 and greater quakes. They were mainly centred in California and one cluster in Mount St Helens in Washington State. Four of them were in Northern and Central California.

Altogether, there were seven straight line corridors (SLCs) aligned with the Sumatran earthquake epicentre. What appears to be local earth tremors in California seem to have a major impact on the other side of the world.

The contributing reciprocal events leading to the major quake in Sumatra seem to be coming from California. But it should be noted that it is not the Californian tremors that caused the Indonesia quake. The Californian tremors acted only as the indicator or footprints of the passing energy beam sweep. Follow these footprints, and it will show us where the culprit is heading!

As stated earlier, earthquakes seem to occur in a reciprocal manner. For example, if the tremors clustered at one end of the corridor, then its opposite side will also receive a fair share. What happens in California will be duplicated in Sumatra, except it is "compacted" to be just one big blow as all the SLCs converged on a single point.

Due to lack of data for below magnitude 2.5 tremors in most areas outside the United States, there were no other observable corridors in alignment with the Sumatran epicentre. Only two other earthquakes in Indonesia were in a corridor with that epicentre, but they occurred earlier on 24 December 2004. It stretched from western Java, passed over the 005853Z epicentre, and ended at the 203809Z epicentre to the north-west of Sumatra.

The Indonesia corridor is shown below in Figure 17.

*Figure 17. The straight line Indonesia corridor with three epicentres.*

The quakes in California are of low magnitudes because the corridors and clustering of events were spread out and hence less stressful. But the opposite situation was encountered at the Sumatran epicentre. As all of them passed over the Sumatran fault line, a major stress was induced, and at 005853Z on 26 December 2004, due to the buildup of stress, gravity lost its grip and the ground jolted in a violent quake.

Figure 18 below shows the alignments of seven US corridors terminating at the Sumatran epicentre. The corridors narrowed down over Sumatra to become a one big punch.

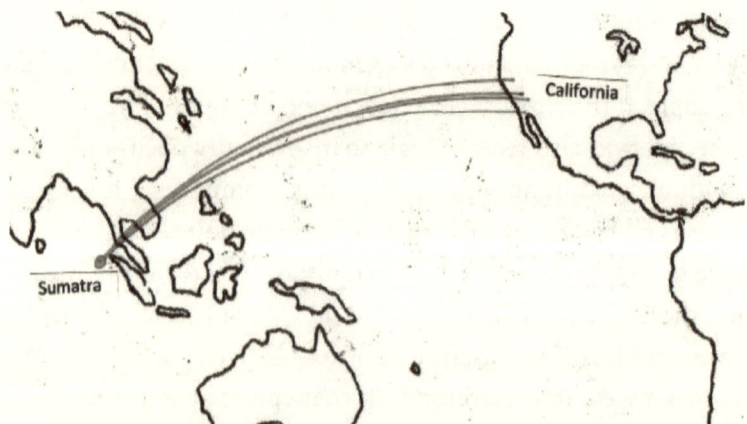

*Figure 18. The orientations of the seven corridors originating from California and Mexico terminated at the Sumatran epicentre. Only four were visible because the other three were too close together.*

Let us now take a look at the Great Sichuan or Wenchuan earthquake in China on 12 May 2008. Will it show the same pattern as seen in the Sumatran earthquake? The Chinese quake measured at magnitude 8.0. Over 69,000 people lost their lives and left nearly 5 million people homeless.

We will present here only the spatial tremor distribution across the globe for a twenty-four- to forty-eight-hour period before the quake strike. To monitor for any big quake, we should look at a seven-day period. The important element that we should looked for is the presence of the straight line corridors (SLCs). That is the best indicators to predict where a major quake will strike.

There were several SLCs during the period, but the significant ones are where most of them converged or passed over the Sichuan epicentre. The scenario should be similar to the one seen for the 2004 Sumatran earthquake. In fact, all big quakes will show the same pattern.

Table 6 lists five straight line corridors, four originating from Greece and one from the Hindu Kush region in Afghanistan. The events started from 000642Z on 11 May 2008, till 062801Z on 12 May 2008.

Table 6. The straight line corridors from Greece and Hindu Kush on 11 May 2008.

| Corridor | Epicentre in Corridor | Dev (km) | Epicentre in Corridor | Dev. (km) | Epicentre in Corridor | Dev. (km) | Length of Corridor (km) | Orientation (deg) |
|---|---|---|---|---|---|---|---|---|
| 005415Z – 062801Z | 005415Z S. Greece | 8.49 | 210935Z S. Greece | 0.00 | 035956Z Greece | 3.53 | 7,317 | 68.50 |
| 205600Z – 062801Z | 205600Z S. Greece | 3.82 | 075735Z S. Greece | 1.06 | 060126Z E. Sichuan | 2.76 | 7,333 | 68.20 |
| 011613Z – 062801Z | 011613Z Crete, Greece | 0.00 | 033356Z Crete, Greece | 1.21 | 062801Z E. Sichuan | 0.00 | 7,075 | 69.22 |
| 070202Z – 062801Z | 060126Z S. Greece | 7.32 | 075735Z S. Greece | 8.69 | 205600Z S. Greece | 7.14 | 7,320 | 68.31 |
| 155941Z – 062801Z | 024219Z Hindu Kush | 3.43 | 205708Z Hindu Kush | 5.03 | 062801Z E. Sichuan | 0.00 | 3,110 | 91.50 |

The known corridors were highlighted in Figure 19 below. The SLC orientations were all convergent to one single point (i.e., the epicentre of the Sichuan earthquake).

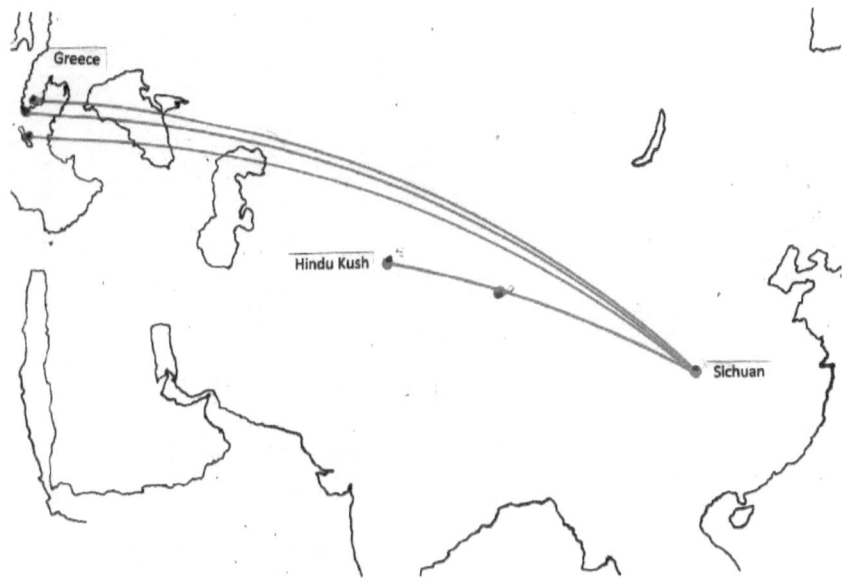

*Figure 19. Four corridors that probably induced stress and resulted in the Great Sichuan Earthquake of 2008.*

On 11 and 12 May 2008, there were not less than sixteen tremors being recorded in Greece, with the most of them located in the southern part of the country. However, the tremors are only for magnitude 2.5 or greater. This means that there are many more low-magnitude quakes that were not documented in the USGS Catalog. These low-magnitude tremor data, when available, could further strengthen the validity of these straight line corridors.

One other corridor in alignment with Sichuan is from the Hindu Kush region in Afghanistan. There were two pockets of seismic activities: one is in the Hindu Kush, and the other is along the way at the Xinjiang-Xizang border region.

California, with its numerous low-magnitude quakes, also contributed at least four SLCs towards Sichuan. Two other SLCs from the United States originated from Utah and Nevada. These corridors are shown in Figures 20 and 21 below.

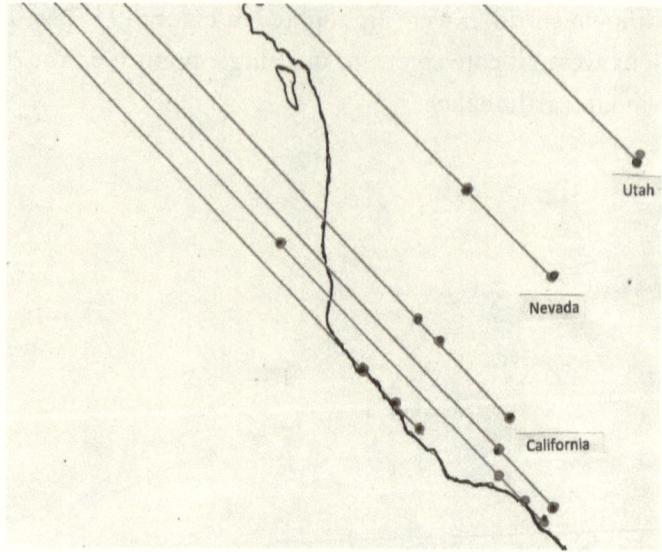

*Figure 20. Six SLCs converging towards Sichuan, China, on 11 and 12 May 2008.*

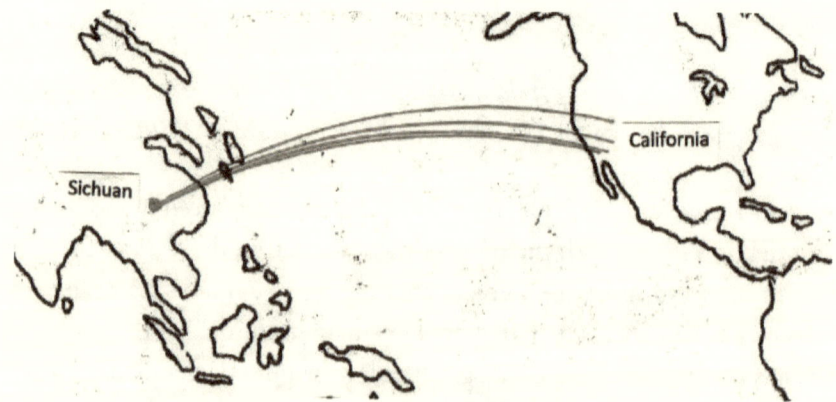

*Figure 21. The United States SLCs contributing to the gravitational stress at Sichuan, China.*

We can say that the stress induced at Sichuan was the direct result of the passage of the energy beam sweep from Source X. This is indicated by the presence of these corridors focusing their bearing and ending at Sichuan as a single point. To use an analogy, it is like if one aircraft dropped just one bomb on a target, the damage inflicted was minimal. On the other hand, if several aircraft were to hit the same target, the results would be devastating. The fault line at Sichuan also broke down with the constant gravitational stress induced during a twenty-four-hour period.

The absence of low-magnitude tremors (below 2.5) in the USGS Catalog from other parts of the globe has prevented us from seeing the complete picture of whether there existed other SLCs leading towards Sichuan. It would be painstaking and time-consuming to obtain low-magnitude earthquake data from the various seismological services around the world.

Before we close this chapter, let us take a look at another great earthquake. It occurred on 16 January 1995, at Kobe, Japan, and is also known as the Great Hanshin earthquake. The quake struck at 204652Z, measuring a 6.9 magnitude with its epicentre located near the south coast of Honshu, Japan, or 20 kilometres from the city centre of Kobe. Over six thousand people lost their lives.

There were no fewer than seven SLCs originating from California to Kobe, Japan. Another corridor came from Chile. In the Region Metropolitana in Chile on 15–16 January 1995, there were a cluster of earth tremors in a SLC. These localised tremors were telling us something, but no one would have suspected that it would have anything to do with any quakes in Japan.

Figure 22 below shows the eight SLCs converging to one location—the epicentre where the Great Hanshin earthquake occurred.

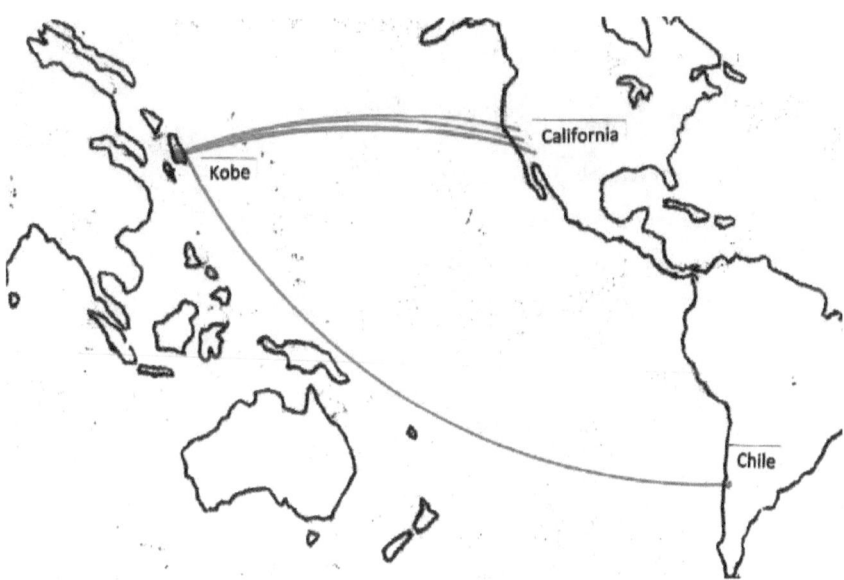

*Figure 22. Californian SLCs and one from Chile converged over Kobe, Japan, on 16 January 1995.*

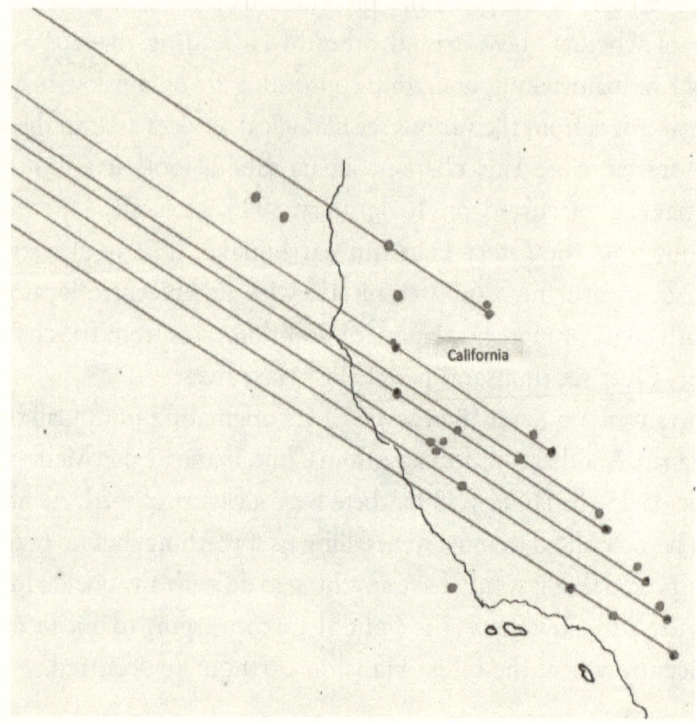

*Figure 23. Seven SLCs in California with orientations towards Kobe, Japan.*

In Table 7 are eight SLCs with a width of less than 5 kilometres. There were no fewer than seventeen earth tremors in the Metropolitana region in Chile in a forty-eight-hour period before the major quake struck Kobe. There were also numerous tremors in California in a twenty-four-hour period before the stress built up in Kobe, Japan.

The Chilean and Californian SLCs are the indicator or footprints to the direction of the energy beam sweep from Source X. To predict any major quakes, there is an urgent need to monitor the SLCs and see in which direction they are heading. The point where all the SLCs crossed is the location where a major quake will strike. Watch these lines (SLCs) converge, and thousands of lives could be saved.

Table 7. *The SLCs leading to the Kobe, Japan, earthquake on 16 January 1995.*

| Corridor | Epicentre in Corridor | Dev. (km) | Epicentre in Corridor | Dev. (km) | Epicentre in Corridor | Dev. (km) | Length of Corridor (km) | Orientation (deg) |
|---|---|---|---|---|---|---|---|---|
| 215437Z (Chile) – 204652Z (Japan) | 011703Z Metropolitana, Chile | 1.57 | 074705Z Metropolitana, Chile | 0.52 | 201220Z Metropolitana, Chile | 0.63 | 17,745 | 278.53 |
| | 0816072 Metropolitana, Chile | 1.43 | 194650Z Metropolitana, Chile | 2.18 | 013734Z Metropolitana, Chile | 0.24 | | |
| | 004752Z Metropolitana, Chile | 4.72 | 190431Z Metropolitana, Chile | 2.14 | 080643Z Metropolitana, Chile | 4.26 | | |
| | 192551Z Metropolitana, Chile | 4.54 | 204652Z Kobe, Japan | 0.00 | | | | |
| 143411Z (California) – 204652Z (Kobe, Japan) | 055831Z Central California | 3.81 | 055311Z Central California | 5.12 | 060940Z Northern California | 3.13 | 8,972 | 306.71 |

| | | | | | | | | |
|---|---|---|---|---|---|---|---|---|
| | 102507Z Northern California | 3.21 | — | | — | | — | |
| 191616Z (California) – 204652Z (Kobe, Japan) | 073032Z Coso Junction, California | 5.59 | 020231Z Northern California | 3.66 | — | — | 9,192 | 307.64 |
| 083548Z (California) – 204652Z (Kobe, Japan) | 055250Z Lucerne Valley, California | 0.22 | 060956Z Kernville, California | 3.77 | 035920Z San Francisco, California | 1.33 | 9,332 | 308.18 |
| | 033151Z San Francisco, California | 1.63 | 035722Z San Francisco, California | 2.09 | 090753Z San Francisco, California | 2.75 | | |
| 142529Z (California) – 204652Z (Kobe, Japan) | 134219Z Central California | 2.09 | 172649Z Northern California | 3.31 | — | — | 9,363 | 308.29 |

| | | | | | | |
|---|---|---|---|---|---|---|
| 001802Z (California) – 204652Z (Kobe, Japan) | 105227Z Tehachapi, California | 3.86 | 204512Z Central California | 1.54 | — | 9,402 | 308.37 |
| 070023Z (California) – 204652Z (Kobe, Japan) | 090911Z Borrego Springs, California | 2.55 | 070402Z Ocotillo Wells, California | 0.34 | 091137Z Anza, California | 0.89 | 9,461 | 308.54 |
| | 053123Z Central California | 2.98 | — | — | — | — | |
| 022205Z (California) – 204652Z (Kobe, Japan) | 175257Z Yorba Linda, California | 0.00 | 121522Z Pacoima, California | 2.23 | 123220Z Pacoima, California | 2.41 | 9,509 | 308.65 |
| | 125517Z Pacoima, California | 3.21 | 135713Z Granada Hills, California | 4.82 | — | — | |

From these major events, we now have a better picture on how a great earthquake occurs, and we get an insight into why a major quake can happen at any specific location. To predict future events, with the current lack of knowledge on the dynamics of Source X, all we need to do is to monitor where all the SLCs converged. The point where all the corridors met is the most likely foci to expect a major quake.

Can we use this simple observation to predict a major quake anywhere on Earth? Yes, we can. Let us take Los Angeles as an example. When can we expect a major earthquake to strike this city and gives its four million inhabitants the shock of a lifetime?

Simply plot the line alignments (SLAs) of all the daily quakes that crossed close to Los Angeles. As a prerequisite, for an earthquake to occur, one SLA from any two earthquakes north and south of Los Angeles must pass over a fault line close to the city. With that, the stress $p1$ has been induced on the San Andreas fault.

Now, all it takes is several other passes to intersect at $p1$. With all these successive gravitational stress imposed on the San Andreas, Los Angeles will be hit by a major quake within a twenty-four-hour period.

This is like if you are in a battle zone and happen to be the bad guy. Each day you watch the bombers fly all over the area, but always away from your location. The chances that you will be bombed next are quite low.

However, as you watch the bombers, they are taking a flight path that is getting closer to your position each day, and eventually the chances are you will be hit soon.

All that needs to be done is monitor whether the lines are getting closer to $p1$. When that happens, the chances are a quake is coming soon, usually within a week. All you need to do is to monitor the SLAs for a one-week period.

If several lines (SLAs) did not cross close to Los Angeles in a one-week period, or maybe only one or two lines crossed near the city of Los Angeles, most of the stress in the area would dissipate within a week. Los Angeles would be safe for the moment.

The most important thing to look for is the clustering of earthquakes in other far-off regions such as Latin America, China, Japan, or South

East Asia. These clusters of tremors are the best indicators or pointers as to where a big quake is going to strike. Watch where the SLCs' compass orientations are leading to and to which point the SLCs converge.

Figure 24 shows the hypothetical situation whereby several energy beams sweep close to the Los Angeles area and can cause a major quake by inducing an unrelenting gravitational stress in a twenty-four- to forty-eight-hour period at the San Andreas fault.

Abnormal surge in seismic activities in Alaska, Europe, Chile, Japan, or China can be a good indicator to see several SLCs converging towards Los Angeles. As a reminder, SLA refers to two earthquake epicentres that occurred one after the other in a compass alignment. It is just a straight line connecting two points or epicentres. SLC refers to several earthquake epicentres in a twenty-four-hour period, distributed along a corridor 5 kilometres wide. SLCs can be used as a primary indicator of where a major quake may strike.

When that happens, a quake similar to or greater than the 1906 San Francisco earthquake could repeat itself close to Los Angeles. How do we save four million lives with the current state of uncertainty regarding where and when the big quake will come? The observations highlighted in this chapter are the best indicator that we have at the moment.

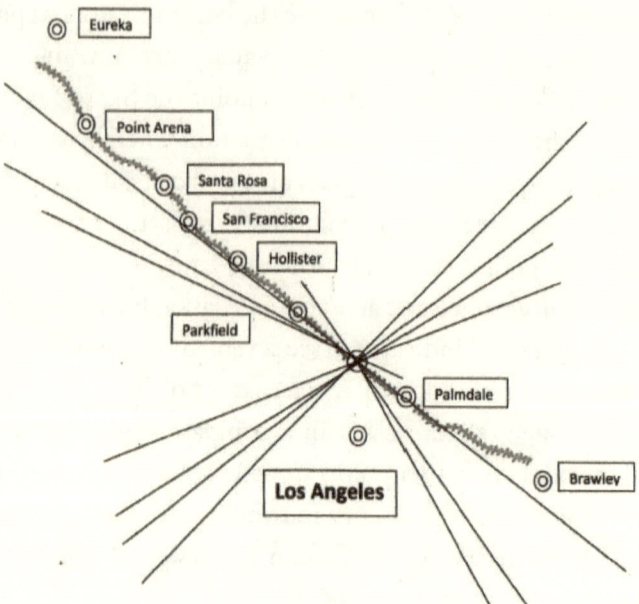

*Figure 24. A hypothetical scenario whereby several SLCs converged over the San Andreas fault line during a twenty-four-hour period.*

Because over a hundred tremors occur each day worldwide, there is always a need for constant monitoring of the SLAs of earthquake epicentres. This type of alignment monitoring can be carried out for all regions in the world because the same principle applies to all regions.

To cause a major earthquake, the energy beam must cross the fault line situated near Los Angeles several times to induce the gravitational stress, which would set the tectonic plates in motion.

Without the actual knowledge of the hypothetical energy beam sweep, at present, to predict the location of an earthquake, we need to monitor the straight line corridors. Plot all the SLCs for any given day and see where they intersect. The location where this happens with be the potential earthquake epicentre twenty-four to seventy-two hours from then.

Remember that for any major earthquake to happen, several SLCs must intersect at one common point. This is where to expect an earthquake measuring 7.0 or greater. The magnitude of the quake will be directly related to the number of SLCs crossing the point. Two intersecting SLCs are also a potential epicentre but usually of low- to medium-magnitude tremors.

In order to make an accurate earthquake prediction, we need to work with the theory of the straight line energy beam sweep. This is the best tool we have at the moment, and that is to watch the SLCs—where they are forming, which direction they are going, and where they will intersect at a common point.

Even if you do not adhere to the theory of Source X in the sky, simply follow this straightforward observation, and you can save millions of lives with a map, a pencil, a ruler, and the latest seismic data. A doctor can save only one life each time on the operating table with all the tools at his or her disposal. Just imagine saving hundreds if not thousands of lives with basic tools—and of course, you need to be a seismologist to be heard!

The bottom line is, the present status quo in earthquake prediction need to change.

## Step 11

# PRELIMINARY PROTOCOL FOR THE PREDICTION OF EARTHQUAKES

We have reached the final part to solve the riddle that has remained elusive for the past one hundred years. This important section outlines a simple method for how to predict earthquakes. It is easy but complex because it involved an unknown source with undetermined properties. To make it controversial, this source has yet to be verified or shown to exist in reality. So how do we predict earthquakes based on an unknown source?

To make a valid earthquake prediction, first we need to understand the mechanics of how earthquake occurs. Let's put the idea of plate tectonics aside for a moment. Take a fresh look from a new perspective. This proposed angle refers to an extrasolar coordinate in the heaven. We will correlate earthquakes with this hypothetical astrophysical source.

The most important element required to make a correct prediction is that the theory to explain the why and how earthquakes occur must be correct. Only a valid theory can gives valid results. Towards this, we need to get to the bottom of the problem.

This boils down to the actual stimulus or trigger factor that induced the Earth to lose its gravitational grip momentarily and caused the Earth to tremble or shake. It is like if you poked a sleeping bear with a stick only once, it would only shudder in response to the outside disturbance. In the case of the Earth, this form of external excitation is what actually causes earthquakes.

Earthquakes are local and transient in nature, but they are interrelated with each other across continents because they shared a common denominator. The presently understood denominator is plate tectonics. This denominator, however, does not work with the prediction protocol presented here. Even without this protocol, the theory of plate tectonics on its own cannot predict earthquakes.

The theory postulates that the plates slide or move against one another, which results in earthquakes. But according to the Source X theory, it is the pinpointed excitation at the fault lines that causes a local momentary gravitational imbalance. This imbalance resulted in a local tremor, just like the bear felt in the area where it was poked.

We assumed that the plates moved to cause earthquakes, but we see that all quakes are confined to a small area marked by a single epicentre for each quake. The other areas of the large plate did not seem to excite any collateral tremors along its large frontal slippage or collision points. If this were to happen, we will see many earthquake epicentres all along the plate front or boundaries at about the same moment. But this was never the case. We see instead isolated earthquakes jumping or occurring across the continents seemingly unrelated to the local plate movements.

The alternating shifts of earthquake epicentre north or south after one another indicate that there is another explanation to it.

If we look closely, the earthquakes appeared to occur in pairs. This pairing characteristic does not correspond to plate collision, abrasion, or sliding. These types of plate movements can happen only when gravity loses its grip. Plate tectonics is therefore only a secondary effect. It is the gravitational excitation that causes the earthquakes.

To achieve any successful earthquake prediction, we need to monitor the rate of rotation and the tilt angle of the hypothetical powerhouse. This can be done only once we can identify and confirm the existence of this source. At present, this is only hypothetical. But the evidence, based on the nature and spatial-temporal distribution of earthquakes, seems to support its possible existence.

Before we go to the prediction protocol, it would be better if we understand the following observed characteristics of earthquakes:

1. Several earthquakes are aligned in straight line corridors (SLCs). The width of these corridors is estimated to be less than 5 kilometres wide.
2. These straight line corridors are usually in a general north-east–south-west or north-west–south-east orientation. Close-proximity earthquakes are always on the north-east–south-west or north-west–south-east orientations. Cross-continents epicentres showed the same trend. This is suggestive that all global earthquakes share a common denominator.
3. Individual earthquakes are paired events sharing the same alignment (SLA). This means that on each SLA, there will be two earthquakes occurring consecutively, one after the other. The distance between two consecutive earthquakes could be far apart or close together occurring as a close proximity double quake.
4. These straight line alignments will shift or flip in its opposite direction every few minutes. The effect is like a seesaw. The holding period of the corridors, or how long each corridor is active, will depend on the rotational rate and energy emission of the hypothetical source. The angle of flip between two corridors is also determined by the spin rate of the source.
5. Earthquakes seem to occur in three different modes: fast, medium, or slow. In the fast mode, there is an earthquake on average every thirty-seven seconds. In the slow mode, on average we can see an earthquake every seventeen minutes. But most of the time, it is about every five minutes. This is the reflection of the behaviour of the trigger source, which could either be rotating or spinning in three different modes (fast, slow, or intermittent). The rate of spin for the source is determined by the amount of energy it releases at its pole(s).

6. The so-called aftershocks are the direct result of the contributing source going into a rapid spin. When it slows down, aftershocks will end and earthquakes will return to the normal tempo of one tremor every five minutes.
7. The optimal striking angle appears to be at the start or end of each beam sweep.
8. The observed tendency is that earthquakes tend to alternately shift the striking location either north or south one after the other, all while maintaining its general north-east–south-west or north-west–south-east orientations.
9. Earthquakes come in two variations. On any given day, earthquake occurrences can either be normal or abnormal. By normal, earth tremor occurs randomly below magnitude 6.0 for a twenty-four-hour period or more. When a major quake of magnitude more than 7.0 occurred, with its foreshocks and aftershocks, then it could be considered abnormal. When the aftershocks subsided or ceased, the situation returned to normal.

With the understanding of how earthquakes behave, we can outline the protocol on the preliminary prediction of earthquakes epicentres. The simple and basic steps are listed below.

**Preliminary Protocol for the Prediction of Earthquakes**
1. Get a big map and mark all the significant or active fault lines
2. Plot the first two earthquakes of the day on the map.
3. Make a straight line joining the two epicentres. This will be your reference line to predict the next quake.
4. This line is called a straight line alignment (SLA). Extend and stretch the line across the map. This is done because other epicentres in the day and the following days will continue to fall on or near this line. When this happen, this line is called a straight line corridor (SLC). The width of this corridor is about 5 kilometres.
5. Once you have this line drawn, two things can happen. (a) This line could flip to its opposite side, taking more than 70

degrees. (b) Its orientation is maintained either in the north-west–south-east or north-east–south-west orientation with small angle shifts.

6. If the first two earthquake epicentre for the day was in the north-west–south-east orientation, shift this line several degrees up or down until it crosses a fault line.
7. The third quake for the day will occur on this new line. If the first quake on the first SLA is in the north and the second recorded quake is in the south, then the third quake will be to the north-east or south-west of the first quake's epicentre. This means that the quake has shifted from the south to the north.
8. Always remember to alternate the position of the quake from north to south and vice versa, depending on which current epicentre location you are working with.
9. However, if the first SLA flipped to its opposite side, then the third quake will be on this new alignment, and the quake will be positioned to the south of the first quake.
10. If you are with the first SLA, which is in the north-west–south-east orientation, and are assuming that the new SLA will flip to its opposite side (i.e., the north-east–south-west orientation), use the average flip angle of 36 degrees from true north as a reference line. Shift this line up or down the 36-degree line until it crosses a fault line. You will have a few potential earthquake epicentres at hand. The third quake will occur at any one of the points where this SLA crosses a fault line.
11. If you are in the north-east–south-west orientation line, it could flip to the opposite side, which is north-west–south-east. When this happens, use 317 degrees as the reference bearing, or 43 degrees left of true north, and shift it up or down until it crosses a fault line. You will find a few potential earthquake epicentres for the next quake.
12. For the fourth quake, it will shift back to the north. Follow this orderly manner, and you should be able to follow the quakes for the rest of the day.

13. This protocol works only for a normal day where there are no major quakes and aftershocks. For an abnormal day with many foreshocks and aftershocks, and of course the main shock, this protocol can still be used, but the area of earthquake concentration will be confined to a specific geographical area with occasional spillover to the other parts of the world.
14. Remember that every two SLAs behave like a seesaw. By shifting the line, you can see that the quakes will take on a new alignment opposite it. This could be in the form of a small angle shift or a large angle flip to the opposite side.
15. When following this protocol, bear in mind that sometimes there are close-proximity double quakes. When this happens, the quake will not occur far away as denoted in this protocol. The quake will strike very close to the first tremor and in the same area.
16. Refer to Figure 25 to get a clearer picture on how earthquakes or their epicentres are distributed with respect to the seesaw effect.

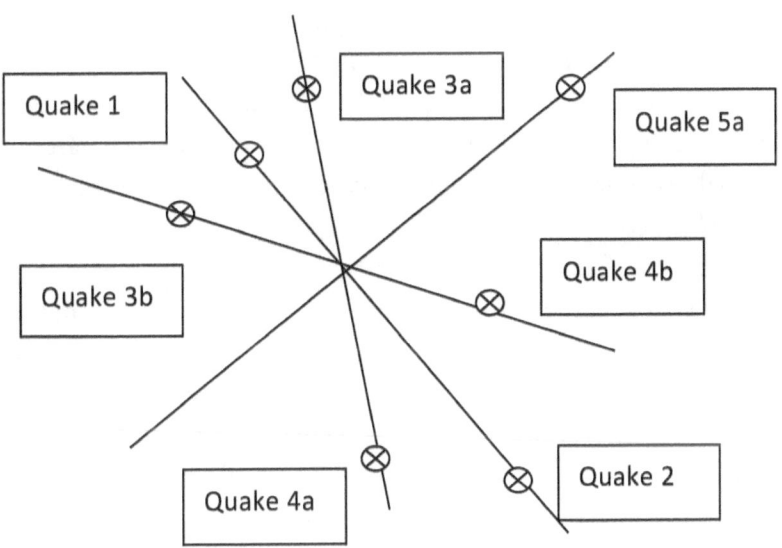

> ***Follow these steps to see where the next earthquake epicentre is located.***
> - Identify the current line alignment for Quake 1 and Quake 2. Use this line as the reference line (SLA).
> - Shift the angle several degrees to the left and right of the line with respect to the nearby fault lines.
> - A tremor will occur close to Quake 1's epicentre, either at Quake 3a or 3b.
> - The fourth quake will strike south of Quake 3a or 3b at Quake 4a or 4b.
> - The fifth quake will be northward of Quake 4a's or 4b's epicentre. It could be on the opposite orientation line at 5a or still holding in the north-west.

*Figure 25: Protocol on where to predict an earthquake epicentre.*

As a reminder, if the general orientation of two quakes is north-west–south-east, then the next quake is expected to be in the north-east or south-west of Quake 1. Identify where the fault lines are located, and your chances of predicting the quake epicentre is better.

If the north-west–south-east orientation flipped to the opposite side, make an estimate of the flip angle. The mean angle is 36 degrees. The third quake should be either to the north-west or south-east of the line. To determine the earthquake epicentre, look where the major fault lines are located.

Due to insufficient earthquake data, especially those measuring 2.5 or lower, we cannot show the precise progression of each tremor in a twenty-four-hour period. But in principle, this is how earthquake can be predicted correctly in term of the location of the epicentre. There is still, however, the uncertainty of the time factor because we do not know how fast Source X is rotating or spinning. This will affect its orientation. However, we can still use the three time frames of thirty-seven seconds, five minutes, or seventeen minutes to expect a quake.

In principle, earthquakes will occur one after the other, as shown in Figure 26.

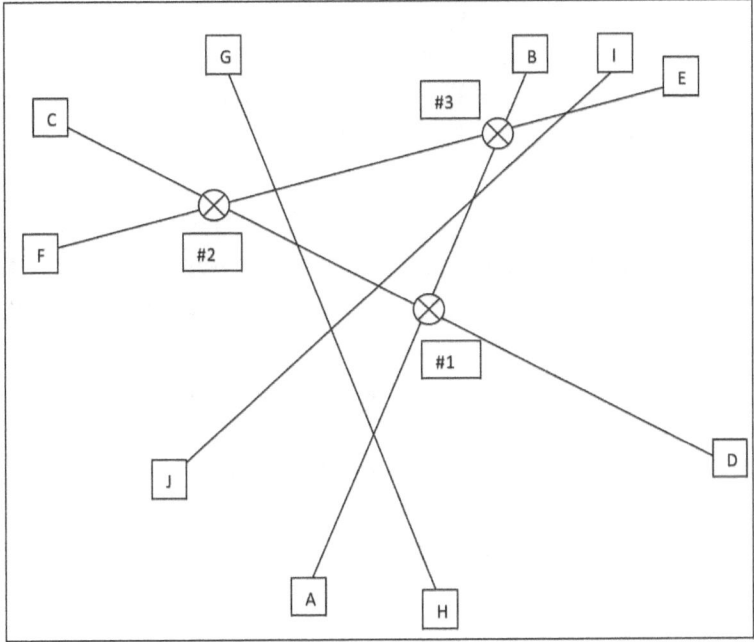

*Figure 26. Distribution pattern of earthquake epicentres in space with respect to the changing orientation lines.*

1. Assume that the first straight line alignment (SLA), or in reality the first beam sweep of the day, is line A to B. Any fault lines in its path will receive a gravitational stress *p1*. A low-magnitude tremor could be felt anywhere along this line where there is a fault line present.
2. After a certain time period—say, between seconds to minutes—the beam sweep flips to its opposite side and takes a new orientation, C to D. We can expect an earth tremor to occur at point 1 because the stress has become *p2*.
3. After a certain time frame, the beam sweep shifted its orientation again with respect to true north and assumed a new SLA, E to F. The next quake in the region will be at the cross point of lines C-D and E-F. But this new SLA will also cross an earlier line A-B, so we can expect an earthquake also at point 3. We can determine the distance between succeeding quakes by looking at where the SLA crossed each other.

4. This pattern will follow true for the rest of the day and will continue to occur for all other days on the same principle. For the crossing points where each SLA crossed, if there is no quake in a twenty-four-hour period, it will occur later, between forty-eight and seventy-two hours or even longer.
5. If we have the earthquake data for all magnitudes, especially those greater than magnitude 1.0, it is possible to predict the earthquake epicentre right on the dot where two lines of the beam sweep crossed at a fault line.
6. Plot all the SLAs and any fault line that it crossed. You will see an earthquake within a one-week time frame. The epicentre of the quake will be smack on the foci where two SLAs crossed.

To recap, this handbook proposes that there exists an astrophysical powerhouse outside Earth that emits gravity waves in a very pinpointed and directional manner. The effect is like a surgical knife that cuts across the face of the Earth in a straight line.

The fault lines are like an open wound that are most susceptible and will feel the most impact when exposed to an outside interference, such as the excitation from energetic gravity waves. This type of gravity wave cannot be coming from faraway sources like black holes, neutron stars, quasars, or any planets with large enough mass to exert any significant gravitational influence on Earth.

At present, the only nearby source, the sun, has a large mass and gravity, but it is not enough to trigger earthquakes in a pinpointed manner. There must be some kind of pulsating or beaming astrophysical powerhouse nearby to cause earthquakes on Earth and all the nearby planets.

The theory of Source X can explain the how and why of earthquake occurrences in a better light than the theory of plate tectonics.

The earthquake prediction protocol presented in this chapter is still in its crude form. However, it can predict earthquake in a slightly better position than the current one, which stood at nil.

This handbook is just an introduction and suggestion to the possible existence of an astrophysical powerhouse. If the assumed behaviour of this source is correct, then it should complement the observed characteristics of earthquakes on Earth in time and space. In other words, if the theory is correct, then all the global earthquakes will be where they should be in time and space.

With these in mind, now we have an almost complete picture (but not yet accurate) on how and where the earthquake will strike. Our only remaining problem is that we are still uncertain on the speed of the beam sweep and when it will cross point $p1$ to determine when the next quake will occur. For this, we need to rely solely on the average time of global earthquake occurrence.

If the rotational speed of the source is fast, we can expect the succeeding quake to occur on average within thirty-seven seconds. At its normal rate, it would be within five minutes. If it slowed down, it will occur around seventeen minutes.

So there we are. To predict earthquakes, we now know the general area to expect the tremor—with several possible locations on hand. We also know when it is striking, and we have three time periods to choose: thirty-seven seconds, five minutes, or seventeen minutes. All three time periods, however, do not provide ample time to raise a warning signal.

This observation, if it turns out to be a valid one, will make earthquake prediction something very close to the truth for the first time in history. Because the data to support the observation comes from only a small number of earth tremors on one date, we need to check for its validity by looking at another twenty-four-hour period's data. This is presented in the next chapter.

If the theory presented is correct, then all earthquakes will follow the same distribution pattern.

To make any accurate earthquake predictions, it is important that we identify the stimulating source and understand its characteristics. With proper computation, predicting earthquakes will be much more accurate than forecasting the weather.

At this early stage, you may not be able get it right for each succeeding quake in time and space. This is probably due to the uncanny behaviour of Source X. Once we can locate and identify Source X and study its behaviour, this preliminary prediction protocol is of no more value. Instrumentation and computation will take over.

Step 12

# CROSSCHECK THE VALIDITY OF THE PREDICTION PROTOCOL

We have come to the final chapter of this handbook. What we need to do next is to check the validity of the earthquake prediction protocol. If it is valid, then all the earthquake epicentres will be where they should be in time and place.

In other words, there is a predetermined mechanic behind it. If the patterns highlighted in this handbook are valid, then earthquakes are predictable.

To check the validity of the observations and findings in this book, we have taken another date of a normal earthquake day and follow its progression in time and space for a limited period, on 29 April 2016. All the earthquake data for the day from the USGS Earthquake Catalog is presented in Appendix A.

Let us follow the events of the day by taking the first two quakes as our reference point or reference line (SLA). We will use graphic illustrations to get a clear picture of where the earthquake epicentres are distributed and where the predicted quake will occur next.

Let us take a look at the first few seismic events of the day and see how the picture builds. From here, you can get the idea of how the quake occurred in any given area. The orientation lines between two succeeding quakes determine the straight line alignment. This alignment is the most important indicator for the next possible earthquake epicentre.

But before any quake can occur, the fault line concerned must be given the first initial stress, called *p1*. When the SLA crosses the point *p1* for the second time, only then can we expect to see an earthquake as the stress induced has become *p2*.

It should be kept in mind that when we mentioned SLAs, it is actually the footprints of the passage of the gravity wave beam across the face of the Earth. This is the energy that induces the ground stress. We are working on the earthquake epicentres, or the foci where this energy beam triggers an earthquake, as the reference points to predict the next quake.

The first recorded quake for 29 April 2016, was a magnitude 1.42 tremor that struck The Geysers in California at 001747Z. The second quake of the day hit Alaska nine minutes later at 002651Z up north. By joining these two epicentres, we get the SLA, or orientation line, of 334.09 degrees.

We see that the first epicentre of the day was in California. The second tremor moved north to strike Alaska. As stated earlier, the location of the quakes tend to alternate each other in their spatial orientation.

The California-Alaska SLA will shift its orientation due to the seesaw effect. With the new alignment, the next quake (the third one for the day) is expected to shift back to the south. So where would the third quake be?

Figure 27 below gives the general picture how the California-Alaska SLA shifted westward from the Alaskan epicentre to create a new alignment, which will put Perry, Oklahoma, in its line of sight.

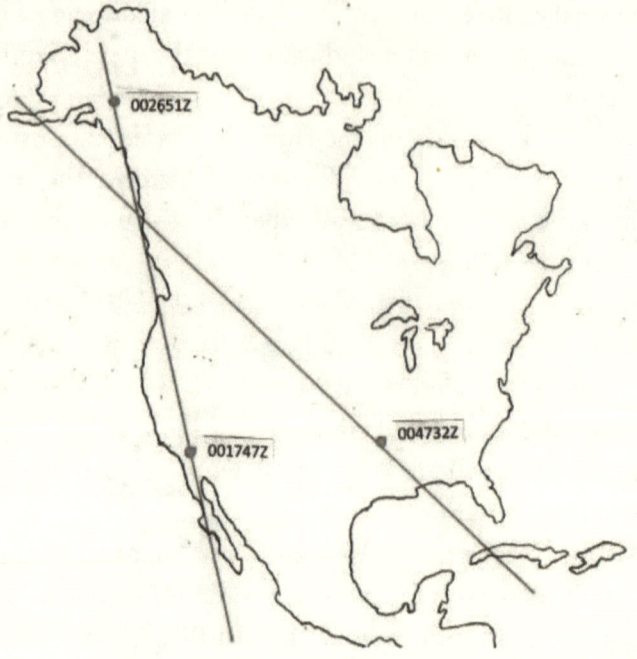

*Figure 27. The SLAs for the first three earthquake epicentres on 29 April 2016.*

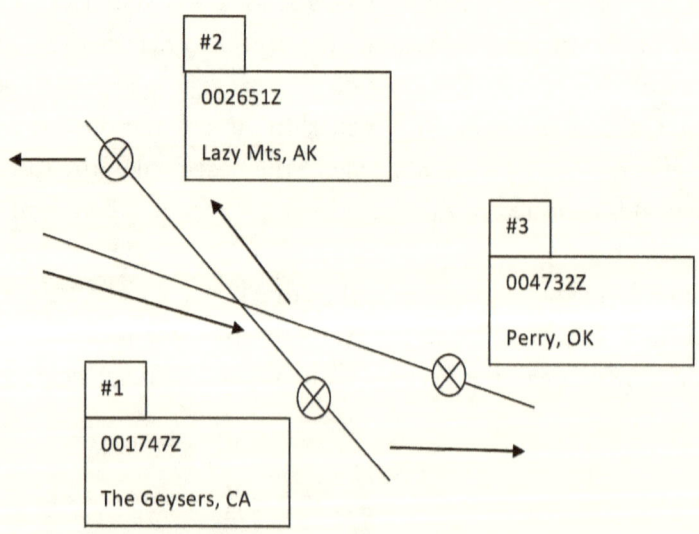

The figure above showed the line shifted to the left, or more specifically the point 002651Z in Alaska moved westward. This westward shift creates a new straight line alignment. The third quake for the day should be on this alignment and to the south of Alaska. Twenty-one minutes later, a magnitude 2.5 tremor struck 8 kilometres west-south-west of Perry, Oklahoma.

After the 004732Z Perry earth tremor, we can expect the quake to shift back north. But the line orientation flipped to the opposite side. This new line alignment became north-east–south west. This drastically changed the location and orientation of the next quake. It did not strike north but to the south. A quake, three minutes later, struck far to the south in Papua New Guinea at 005019Z.

According to the protocol, either we can expect the line orientation to flip back to the north-west–south-east, or there could be a slight angle shift from the 004732Z–005019Z SLA. A quake occurred still farther south registering a magnitude 4.8, 40 kilometres west-south-west of Vanuatu at 005247Z. This showed that the SLA has flipped back to the north-west.

The seesaw effect is again demonstrated here as the line from Perry shifted westward, resulting in the new earthquake epicentre in Vanuatu at 005247Z. This epicentre is located to the south-east of the earlier Papua New Guinea quake at 005019Z.

Figure 28 below shows the two SLAs between Perry, Oklahoma, and the quake in Papua New Guinea. The westward shifted SLA resulted in a quake in Vanuatu.

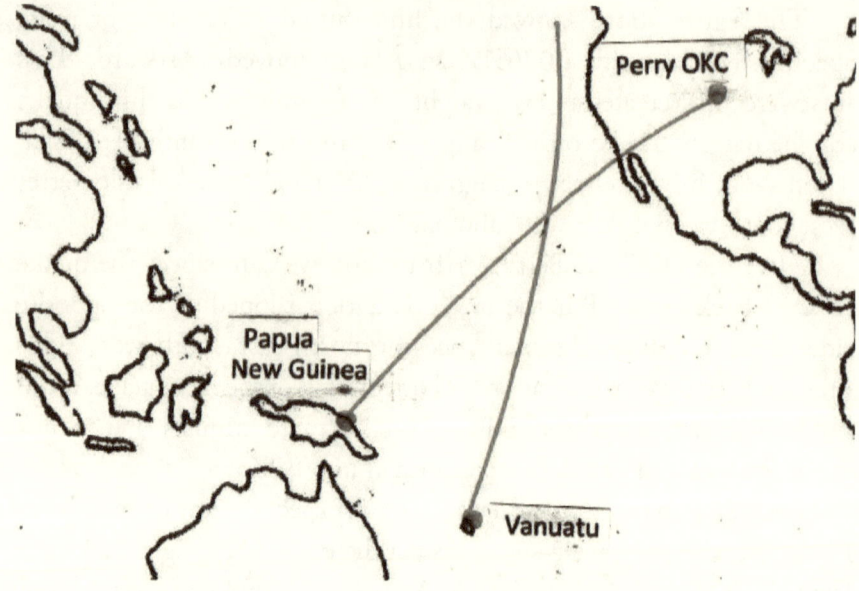

*Figure 28. The seesaw effect of the Perry–Papua New Guinea tremors and the earthquake in Vanuatu.*

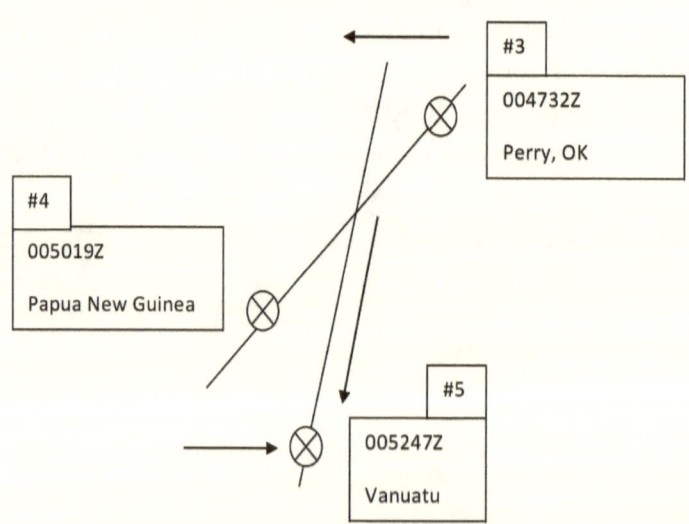

After Vanuatu, where will the next quake occur? As usual the Vanuatu SLA will shift westward. This new orientation will put Alaska,

California, Mexico, or the northern parts of Latin America in line for the next quake.

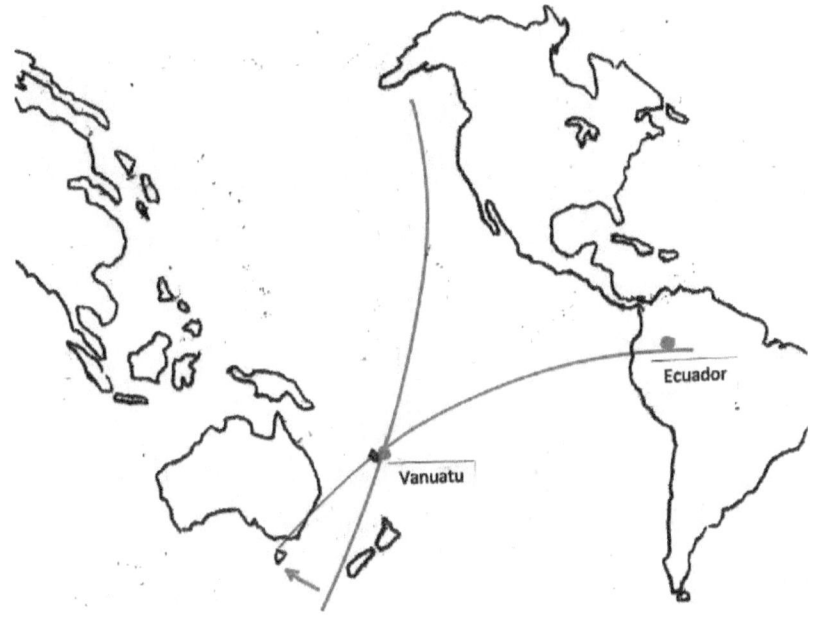

*Figure 29. The Vanuatu SLA shifted westward, resulting in a big flip to the opposite side.*

About five minutes later, as the Vanuatu SLA shifted westward, its opposite end moved down like a seesaw. This shift in line orientation resulted in the quake striking Ecuador at 005654Z.

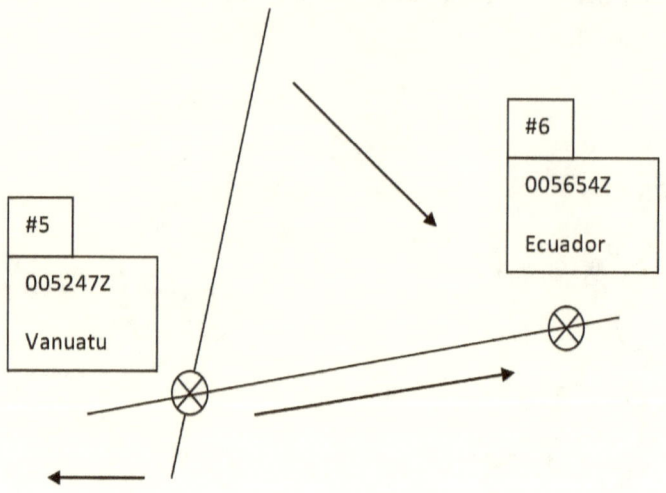

Just two minutes later, the line orientation flipped to the north-west. This new orientation resulted in the Montana earthquake at 005847Z. And three minutes later, the Ecuador-Montana SLA flipped to the north-east, and this new alignment placed Vanuatu back in the striking position. A quake occurred there again at 010153Z.

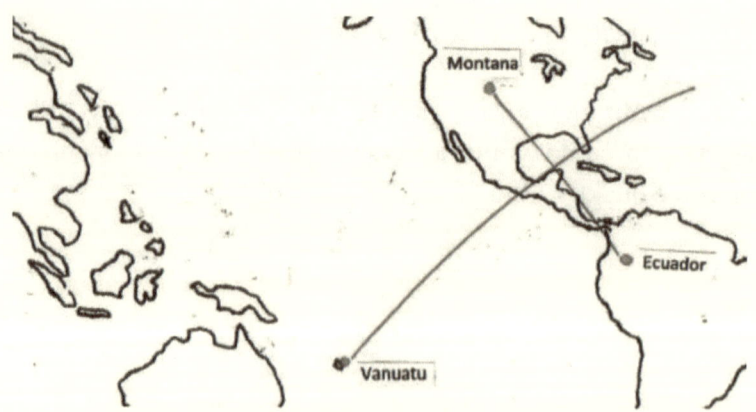

*Figure 30. The Ecuador-Montana SLA flipping to the north-east, resulting in the second Vanuatu quake.*

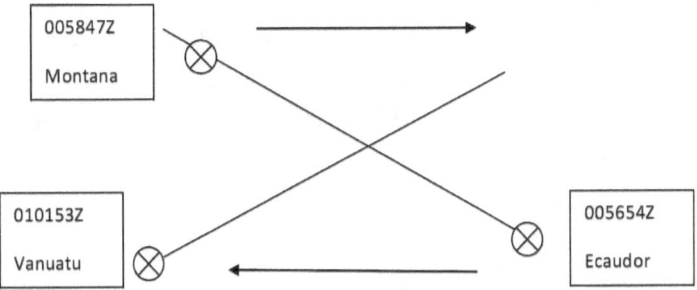

During the two to three minutes after the Vanuatu quake, the optimal striking point shifted back to the north with a close-proximity double quake in Helena West Side, Montana, 29 kilometres apart at 010358Z and 010431Z, respectively.

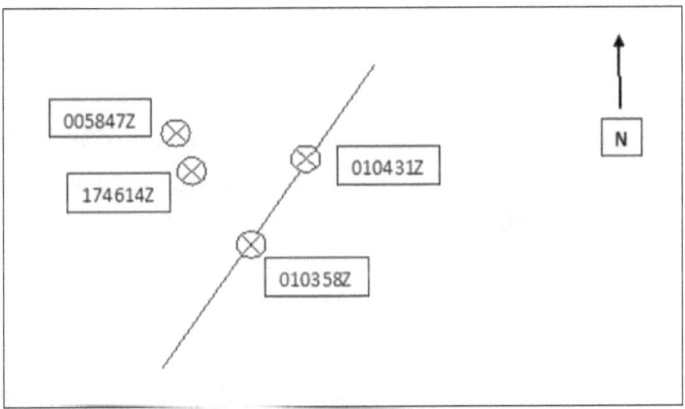

*Figure 31. Close-proximity double quake near Helena West Side, Montana.*

By now you should get the picture that when one quake occurs in the south, the next one will be in the north. As to where the next quake will occur, just follow the seesaw effect. Close proximity double quakes are arranged in the same manner except they are much closer. Whether the epicentre is in the north or the south, always shift the epicentre point westward. But this is only for cases with small angle changes.

When there is a large change, it is a flip, so it will be on its opposite side. That new orientation will be the new SLA, where an earthquake will strike opposite the latest recorded tremor on that line.

After the close-proximity double quake in Montana, where can we expect the next quake to strike?

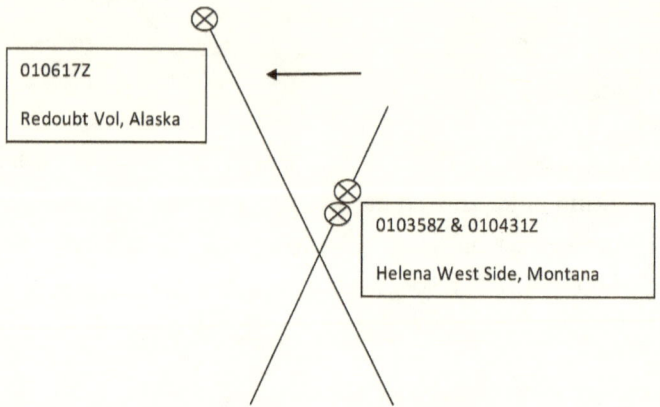

As usual, use the stretched-out Montana SLA for the 010358Z–010431Z tremors and follow the seesaw effect. Shift the 010431Z epicentre point westward, and you will get a new orientation SLA that run north-west to south-east. The next quake will be northward if we take Montana as its southern counterpart.

At 010617Z, a magnitude 1.7 quake hit 46 kilometres from Redoubt Volcano, Alaska. Again we see that the quake shifted to its opposite north.

After the Redoubt Volcano tremor, by now you should be able to make a educated guess that the next quake should occur in the south. At which location would it be in the south? Remember to always use the seesaw effect to see where the next quake is likely to strike. However, we have a small problem because we are not sure how wide an angle this seesaw will shift west from the Montana-Alaska SLA.

For the next quake, the angle shift was quite small, and a quake struck in the south at 010636Z, 11 kilometres west of Caldwell, Kansas. Refer to Figure 32 below.

*Figure 32. The SLAs for the Montana, Alaska, and Kansas earthquakes.*

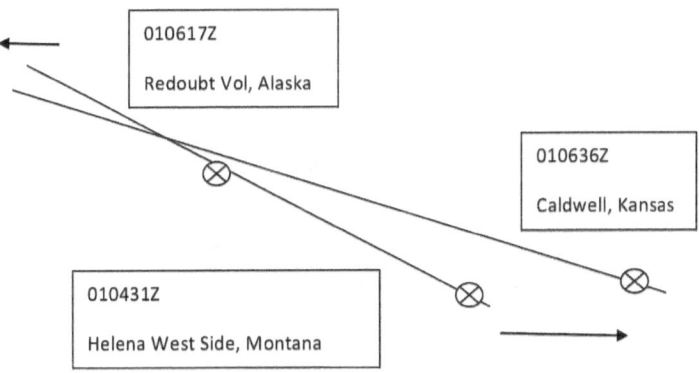

As usual, after Kansas the SLA will shift or flip to its opposite side. This new SLA orientation will place the next quake to the south and in the Pacific Ocean! At 013338Z, a magnitude 6.6 quake hit the northern East Pacific Rise in the ocean. Please refer to the figure below.

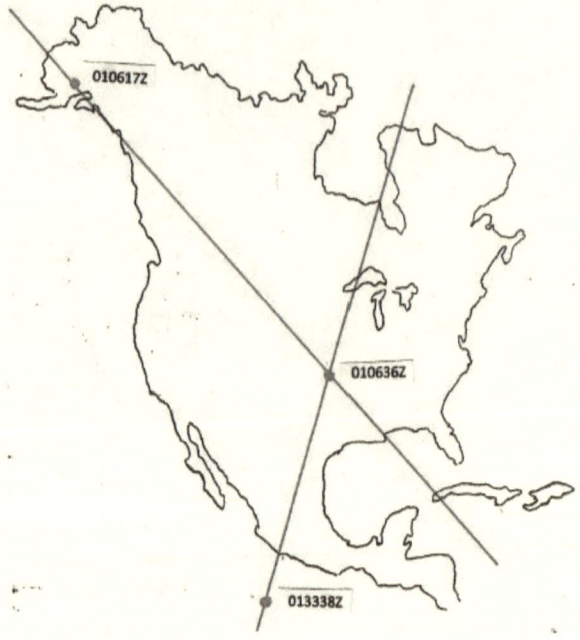

*Figure 33. The extended Alaska-Kansas SLA flipping to the north-east with a quake in the Pacific Ocean.*

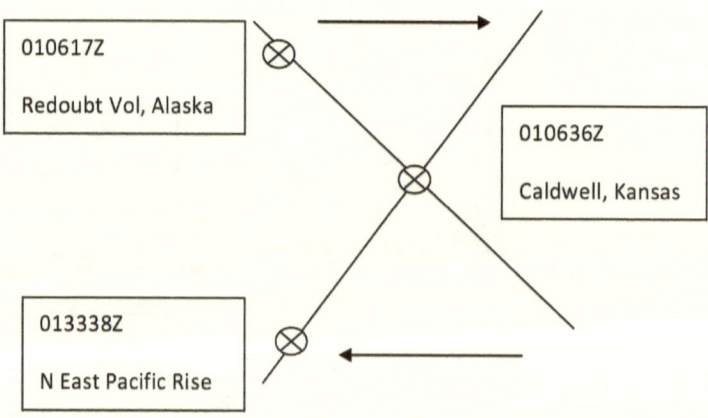

We can see the same picture in the earthquake epicentre distribution pattern for the rest of the day. The only problem is that we do not know the distance between the latest quake and the impending one. But if we know which line alignment, all that needs to be done is to monitor the

SLA crossing which fault line. The distance between two epicentres is determined by the orientation of the new SLA, at which point it will cross the *p1* foci.

SLA will determine correctly were the beam sweep crossed any particular fault line along its track. To make an accurate earthquake prediction, this SLA is the most important element to watch for.

To recap, we have taken a look only at the first six SLAs of the day in this chapter. They are as follows:

1. The Geysers in California to Lazy Mountain in Alaska (001747Z–002651Z)—9 minutes
2. Perry, Oklahoma, to Papua New Guinea (004732Z–005019Z)—3 minutes
3. Vanuatu to Ecuador (005247Z – 005654Z)—4 minutes
4. Helena West Side, Montana, to Vanuatu (005847Z–010153Z)—3 minutes
5. The Helena West Side SLA – (010358Z–010431Z)—1 minute
6. Redoubt Volcano, Alaska, to Caldwell, Kansas (010617Z–010636Z)—19 minutes

Let us take a look again what transpired during the early hours of 29 April 2016, across the globe.

In a time frame of only thirty-eight minutes, the SLAs had their orientations changed six times. During the first nine minutes, that has induced the stress *p1* to two other points on the fault lines, from The Geysers, California, to Lazy Mountain, Alaska. The quakes came later at 064700Z and 103312Z, both in The Geysers, California.

The quake at 064700Z in The Geysers occurred as a result of a new SLA originating from Peru at 063319Z. The stress *p2* was induced at the 064700Z fault line with the passage of the Peruvian SLA. This resulted in the 1.69 magnitude earthquake.

This Peruvian SLA also induced the stress *p1* at two other locations. One was 103312Z, 6 kilometres north-west of The Geysers, which is right on the SLA line. The other earmarked location, also in the same area but 0.24 kilometres away from this SLA, was at the 113018Z epicentre. The latter tremor came about five hours later from a new SLA from the direction of Fillmore, California.

The 103312Z earthquake at The Geysers came from a SLA from the direction of Lone Pine, California.

From these small samples, we can see that for any earthquake to happen, a SLA must pass over the area to induce the stress *p1*. The quake will occur during the second pass, usually from a new SLA orientation.

To predict any earthquake at any fault line, we must monitor the SLAs and see which fault line it passes. Then we can assign the point where the SLA crossed the fault line as *p1*. If another SLA appears later on and crosses the designated *p1* point, a quake will strike precisely at that point. This is how we can predict the location or the epicentre of any quake.

The SLAs do not start or end at two earthquake epicentres; in reality, they actually circumscribe the face of the earth. Any other areas where this great circle line crossed will also received the stress *p1*. For this reason, we can see widely separated quakes occurring less than a few seconds or minutes apart, one after the other across the continents.

What do the observations tells us? We can deduce from them the following points:

1. Once we have identified one SLA, two possible things can happen. The next new SLA would be a shift westward from the latest quake epicentre with a minor angle change, or it could flip to its opposite side with a major angle change.
2. Let us look at the westward shift first. When this happens, imagine it is like a seesaw. When one side moved in one direction, the other side will move in the opposite direction. Because we have one SLA, the new SLA will be a westward shift from that latest earthquake epicentre. If the latest epicentre (reference epicentre) is in the north, the new line shift will place the impending quake epicentre to its south, or more specifically to the south-east, if the reference epicentre is in the north-west. If the reference epicentre is in the north-east, the predicted quake will be in the south-west.

3. We can see two SLAs crossing each other not in the very centre like a seesaw pivot point, but usually offset far from the centre.
4. In the case of a flip from one SLA to form a new SLA, the angle change is large. In the new orientation, the distribution pattern of the impending quake will be the same as in the westward shift case.

To sum up, this tells us the following facts:
1. Earthquakes can be predicted. The spatial-temporal distributions of earthquake epicentres abide to a certain pattern. Epicentres are aligned in a straight line (SLA), and this line changes its orientation on average every five to ten minutes.
2. This earthquake prediction protocol, however, is still in its crude form. It cannot predict all the quakes in a given time frame, but it showed that tremors will occur at the predicted points usually within minutes. Overall, this protocol is still helpful to predict earthquake when compared to our present earthquake forecasting ability, which is nonexistent.
3. All earthquakes (except from mining activities, quarry blasts, ice quakes, and nuclear explosions) shared a common denominator. This can only happen if a single common stimulus that is located outside the Earth acted as the primary triggering earthquake factor.
4. This stimulus and its nature is still unidentified, but it is believed to be an astrophysical powerhouse (or a mass of compact energy).
5. If this stimulus existed in reality, tremors could be induced not only on Earth but also on the moon, the sun, and all the planets in this solar system.

This handbook proposes that all earthquakes were controlled by a common underlying mechanism. This will dictate where and when the earthquake epicentres are set in time and space. The fact that they are aligned on several straight lines or on a great circle denotes a common stimulus that must have its source from none other than the sky.

Earthquakes, therefore, are not random individual events isolated from each other. They are all interrelated and triggered by the same source. With this view, it would be possible to accurately predict earthquakes once we can identify the hypothetical extraterrestrial stimulus.

Any handbook or manual on any matter needs to be tested to see for its merits and validity. Even though the theory presented here is new and controversial and goes against the normal perception of earthquakes mechanics, it is the only one that can predict (albeit on a cat-and-mouse basis at this early stage) an earthquake epicentre that is correct in a specific area and in a time frame of less than twenty minutes.

No earthquake manual (if there is any) in the market today can give such a promising outcome. With this new approach and understanding of the stimulus, we can make a successful prediction to the time, place, magnitude, and depth of the quake once we have studied Source X. Seismology certainly will see a brighter future.

# REFERENCES

Carmenzind, M. "Compact Objects in Astrophysics: White Dwarfs, Neutron Stars and Black Holes." *Springer Science & Business Media* (February 2007), 265.

Evison, F. "On the Existence of Earthquake Precursors." *Annali Di Geofisica* 42, no. 5 (October 1999). 763-770.

Flandern, V. T. *Dark Matter, Missing Planets & New Comets: Paradoxes Resolved, Origins Illuminated.* Berkeley, California, North Atlantic Books, 1993.

Fuqiong, et al. "Studies on Earthquake Precursors in China: A Review for Recent 50 Years. *Geodesy and Geodynamics* 8, no. 1 (January 2017) 1-12.

Geller, R.J., et al. "Earthquakes Cannot Be Predicted." *Science* 275 (March 1997) 1616-1617.

Gribbin, John R., and Stephan H. Plagemann. *The Jupiter Effect.* New York, Vintage Books, 1975.

Hawking, Stephen. *A Brief History of Time.* New York, Bantam Books, 1989.

Hainzl, S. "Seismicity Patterns of Earthquake Swarms Due to Fluid Intrusion and Stress Triggering." *Geophysical Journal* 159, no. 3, December 2004, 1090–1096.

Latham, G., et al. "Moonquakes." *Science* 174 (1971), 687–692.

MacCarthy, G. "A Marked Alignment of Earthquake Epicenters in Western North Carolina and Its Tectonic Implications." *Journal of the Elisha Mitchell Scientific Society* 71, no. 2 (November 1956) 763 - 770.

Miller, S. *Earthquake Swarms*, Advances in Geophysics, ScienceDirect.com/Elsevier B.V, 2013, 1-46.

Okal, E., et al. "On the Planetary Theory of Sunspots." *Nature* 253 (1975), 511–513.

USGS Earthquake Catalog. https://prod-earthquake.cr.usgs.gov.

Xue, Y., et al. "Characteristics of Seismic Activity before the $M_s 8.0$ Wenchuan Earthquake." *Earthquake Science* 22, no. 5 (October 2009) 519 – 529.

Yamashina, K. and Y. Inoue. "A Doughnut-Shaped Pattern in Seismic Activity Preceding the Shimane Earthquake of 1978." *Nature* 278 (March 1979), 48–50.

# APPENDIX A

Earthquake data for 29 April 2016, as listed in the USGS Earthquake Catalog. Tremors due to quarry blast, mining activities, and icequakes have been omitted.

| Time | Latitude | Longitude | Magnitude | Region |
|---|---|---|---|---|
| 00:17:47.610Z | 38.8115 | -122.811 | 1.42 | 6 km NW of The Geysers, California |
| 00:26:51.000Z | 61.5662 | -147.8712 | 2.1 | 57 km E of Lazy Mountain, Alaska |
| 00:47:32.500Z | 36.2538 | -97.3691 | 2.5 | 8 km WSW of Perry, Oklahoma |
| 00:50:19.610Z | -6.0316 | 147.9791 | 4.1 | 64 km N of Finschhafen, Papua New Guinea |
| 00:52:47.490Z | -16.1872 | 167.0266 | 4.8 | 40 km WSW of Norsup, Vanuatu |
| 00:56:54.920Z | -0.6933 | -76.3772 | 4.0 | 70 km ESE of Puerto Francisco de Orellana, Ecuador |
| 00:58:47.950Z | 46.6753333 | -112.486 | 2.86 | 29 km WNW of Helena West Side, Montana |
| 01:01:53.770Z | -16.2393 | 167.2336 | 4.8 | 23 km SW of Lakatoro, Vanuatu |
| 01:03:58.330Z | 46.6616667 | -112.4838333 | 1.31 | 29 km WNW of Helena West Side, Montana |

| | | | | |
|---|---|---|---|---|
| 01:04:31.420Z | 46.6706667 | -112.4793333 | 2.34 | 29 km WNW of Helena West Side, Montana |
| 01:06:17.000Z | 60.0872 | -152.9748,87 | 1.7 | 46 km SSW of Redoubt Volcano, Alaska |
| 01:06:36.180Z | 37.0493333 | -97.734 | 1.83 | 11 km W of Caldwell, Kansas |
| 01:33:38.910Z | 10.2752 | -103.7363 | 6.6 | Northern East Pacific Rise |
| 01:41:00.880Z | 34.024 | -117.7713333 | 1.29 | 4 km NNW of Chino Hills, CA |
| 01:50:46.960Z | -1.4911 | -80.8167 | 4.3 | 31 km SW of Jipijapa, Ecuador |
| 02:04:16.000Z | 61.2955 | -146.7504 | 1.4 | 28 km NW of Valdez, Alaska |
| 02:35:36.000Z | 63.0966 | -151.467 | 1.7 | 110 km NW of Talkeetna, Alaska |
| 02:57:28.000Z | 51.9959 | -173.6749 | 1.7 | 42 km ESE of Atka, Alaska |
| 03:02:45.000Z | 59.284 | -152.6876 | 1.7 | 73 km SW of Anchor Point, Alaska |
| 03:05:28.000Z | 61.5481 | -146.4661 | 1.3 | 46 km N of Valdez, Alaska |
| 03:08:33.000Z | -22.208 | -68.55 | 4.1 | 48 km NE of Calama, Chile |
| 03:19:24.840Z | 34.022 | -118.3486667 | 1.65 | 3 km N of View Park-Windsor Hills, CA |
| 03:20:03.000Z | 51.7079 | -178.4987 | 1.7 | 31 km SW of Tanaga Volcano, Alaska |
| 03:26:59.360Z | 37.7373333 | -110.6658333 | 2.02 | 70 km S of Hanksville, Utah |
| 03:37:07.000Z | 58.5622 | -142.8057 | 2.6 | 168 km S of Cape Yakataga, Alaska |
| 03:48:50.300Z | 36.8622 | -98.0017 | 3.0 | 24 km WNW of Medford, Oklahoma |
| 03:50:55.010Z | 38.7064 | 20.4963 | 4.3 | 18 km W of Nidri, Greece |
| 04:09:26.000Z | 61.6093 | -146.3968 | 1.1 | 53 km N of Valdez, Alaska |

| | | | | |
|---|---|---|---|---|
| 04:20:56.000Z | 60.9837 | -152.0331 | 1.8 | 52 km NW of Nikiski, Alaska |
| 04:22:47.370Z | -27.2761 | -63.4352 | 4.1 | 34 km SW of El Hoyo, Argentina |
| 04:25:49.000Z | 19.6534 | -65.3629 | 3.6 | 144 km NNE of Vieques, Puerto Rico |
| 04:36:36.290Z | 39.674 | 74.125 | 4.7 | 75 km E of Sary-Tash, Kyrgyzstan |
| 04:41:30.000Z | 51.5659 | -176.3737 | 1.2 | 40 km SSE of Adak, Alaska |
| 04:42:21.000Z | 57.8893 | -156.2904 | 2.3 | 91 km SSE of King Salmon, Alaska |
| 04:47:16.000Z | 58.8296 | -136.9188 | 1.7 | 82 km NW of Gustavus, Alaska |
| 04:51:58.800Z | 36.4439 | -97.1271 | 2.5 | 22 km NE of Perry, Oklahoma |
| 05:06:28.500Z | 35.5277 | -96.7724 | 3.0 | 9 km WNW of Prague, Oklahoma |
| 05:10:33.000Z | 61.4357 | -146.4493 | 1.2 | 34 km N of Valdez, Alaska |
| 05:24:57.900Z | 37.5741667 | -118.8571667 | 1.09 | 13 km SE of Mammoth Lakes, California |
| 05:25:00.000Z | 62.0042 | -149.8674 | 2.1 | 17 km S of Y, Alaska |
| 05:26:20.350Z | 40.6473 | -119.5205 | 1.9 | 14 km WNW of Gerlach-Empire, Nevada |
| 05:27:28.000Z | 61.0231 | -150.3213 | 3.0 | 31 km SW of Anchorage, Alaska |
| 05:34:54.000Z | 63.1493 | -150.9013 | 1.3 | 100 km NNW of Talkeetna, Alaska |
| 05:37:20.000Z | 61.1426 | -150.4381 | 1.3 | 30 km WSW of Anchorage, Alaska |
| 05:40:35.000Z | 61.8174 | -154.2792 | 1.6 | 144 km SSE of McGrath, Alaska |
| 05:42:30.000Z | 53.5392 | -167.1793 | 1.7 | 56 km SW of Unalaska, Alaska |
| 05:56:17.900Z | 19.4962 | -65.319 | 3.2 | 130 km NNE of Vieques, Puerto Rico |

| | | | | |
|---|---|---|---|---|
| 05:58:08.500Z | -23.2875 | -174.0647 | 4.6 | 234 km SSE of Ohonua, Tonga |
| 06:09:35.880Z | 33.2523 | 131.2868 | 4.2 | 12 km E of Tsukawaki, Japan |
| 06:10:43.000Z | 61.2443 | -146.7674 | 1.3 | 25 km WNW of Valdez, Alaska |
| 06:18:03.000Z | 69.4357 | -143.6992 | 2.4 | 77 km S of Kaktovik, Alaska |
| 06:32:50.000Z | 53.5032 | -165.6653 | 1.6 | 70 km SE of Unalaska, Alaska |
| 06:33:19.190Z | -15.4185 | -73.1575 | 4.7 | 79 km ESE of Coracora, Peru |
| 06:47:00.740Z | 38.8178333 | -122.794 | 1.69 | 5 km NW of The Geysers, California |
| 07:00:48.300Z | 34.1343333 | -117.4486667 | 1.1 | 4 km NNE of Fontana, CA |
| 07:02:39.220Z | 35.7465 | -82.5756667 | 1.53 | 5 km NNW of Weaverville, North Carolina |
| 07:05:32.000Z | 51.4841 | -176.6527 | 1.7 | 44 km S of Adak, Alaska |
| 07:11:26.000Z | 60.5019 | -151.3033 | 2.0 | 6 km SSW of Kenai, Alaska |
| 07:12:17.000Z | 60.0838 | -152.608 | 2.6 | 45 km S of Redoubt Volcano, Alaska |
| 07:23:37.100Z | 35.5942 | -97.3794 | 2.7 | 7 km N of Spencer, Oklahoma |
| 07:45:28.100Z | 18.5454 | -66.4882 | 2.3 | 9 km N of Tierras Nuevas Poniente, Puerto Rico |
| 07:48:50.000Z | 61.3509 | -147.6304 | 1.4 | 73 km WNW of Valdez, Alaska |
| 07:59:21.748Z | 40.0755 | -119.6983 | 1.3 | 47 km N of Spanish Springs, Nevada |
| 08:08:05.320Z | 19.3593333 | -155.3023333 | 1.63 | 10 km SW of Volcano, Hawaii |
| 08:17:33.000Z | 58.8989 | -152.3383 | 4.9 | 94 km SSW of Homer, Alaska |

| | | | | |
|---|---|---|---|---|
| 08:17:36.010Z | 60.425 | -152.152 | 3.9 | 33 km ESE of Redoubt Volcano, Alaska |
| 08:45:27.620Z | 38.6861667 | -122.8966667 | 2.02 | 8 km NNW of Healdsburg, California |
| 08:56:24.600Z | 35.748 | -82.577 | 1.81 | 5 km NNW of Weaverville, North Carolina |
| 08:59:06.240Z | 34.009 | -118.3466667 | 1.23 | 2 km N of View Park-Windsor Hills, CA |
| 09:05:17.000Z | 36.468 | -98.7537 | 2.9 | 33 km NW of Fairview, Oklahoma |
| 09:13:31.000Z | 58.8993 | -154.5158 | 3.4 | 97 km SSE of Old Iliamna, Alaska |
| 09:24:55.000Z | 63.1361 | -151.4091 | 1.5 | 112 km NW of Talkeetna, Alaska |
| 09:27:24.490Z | 48.0903333 | -122.6628333 | 1.2 | 7 km ESE of Port Townsend, Washington |
| 09:31:36.120Z | 51.1773 | 179.645 | 4.1 | 86 km S of Semisopochnoi Island, Alaska |
| 10:07:13.000Z | 61.809 | -149.7179 | 1.2 | 18 km ENE of Willow, Alaska |
| 10:18:20.840Z | -16.0517 | 167.3768 | 4.3 | 1 km NNW of Norsup, Vanuatu |
| 10:19:08.690Z | 20.1053333 | -155.5483333 | 2.37 | 8 km WNW of Honoka'a, Hawaii |
| 10:20:55.000Z | 52.8462 | -169.5896 | 1.9 | 49 km WSW of Nikolski, Alaska |
| 10:25:52.660Z | 36.3539 | -117.4192 | 1.2 | 64 km ESE of Lone Pine, California |
| 10:33:12.470Z | 38.8240013 | -122.8075027 | 1.01 | 6 km NW of The Geysers, California |
| 10:35:14.980Z | -15.611 | 167.509 | 4.1 | 37 km ESE of Luganville, Vanuatu |
| 10:47:41.000Z | 58.8639 | -152.4375 | 2.2 | 100 km SSW of Homer, Alaska |
| 10:52:32.000Z | 57.9901 | -155.1701 | 1.4 | 86 km NW of Larsen Bay, Alaska |

| | | | | |
|---|---|---|---|---|
| 10:54:51.000Z | 59.4056 | -151.4916 | 1.6 | 26 km S of Homer, Alaska |
| 10:55:20.360Z | 32.391 | -115.247 | 2.77 | 6 km NW of Delta, B.C., MX |
| 10:55:25.010Z | 32.3876667 | -115.2651667 | 3.02 | 8 km WNW of Delta, B.C., MX |
| 10:57:33.000Z | 19.8135 | -155.4165 | 1.66 | 30 km S of Honoka'a, Hawaii |
| 10:58:35.000Z | 62.6422 | -148.0832 | 1.3 | 94 km SSE of Cantwell, Alaska |
| 11:18:18.110Z | 19.3305 | -155.125 | 2.06 | 16 km SE of Volcano, Hawaii |
| 11:20:15.580Z | 34.4513333 | -118.9438333 | 1.64 | 6 km NNW of Fillmore, CA |
| 11:30:18.570Z | 38.7761667 | -122.7256667 | 1.28 | 2 km E of The Geysers, California |
| 11:32:17.032Z | 41.9778 | -120.1739 | 1.0 | 27 km SSE of Lakeview, Oregon |
| 11:32:24.000Z | 59.8495 | -152.9831 | 1.7 | 65 km W of Anchor Point, Alaska |
| 11:43:04.000Z | 18.8607 | -64.2679 | 2.2 | 61 km NE of Road Town, British Virgin Islands |
| 11:48:25.000Z | 51.6866 | -179.7296 | 2.7 | 54 km ESE of Semisopochnoi Island, Alaska |
| 11:59:50.000Z | 60.3785 | -149.8279 | 1.4 | 33 km NW of Bear Creek, Alaska |
| 11:59:52.776Z | 40.6621 | -119.4945 | 1.2 | 13 km NW of Gerlach-Empire, Nevada |
| 12:04:10.700Z | -6.6119 | 103.8711 | 4.4 | 142 km S of Biha, Indonesia |
| 12:07:17.000Z | 59.8895 | -151.6971 | 1.4 | 14 km NNE of Anchor Point, Alaska |
| 12:17:39.000Z | 64.7385 | -155.1378 | 1.1 | 16 km E of Ruby, Alaska |
| 12:19:42.420Z | 33.1783333 | -115.9763333 | 1.43 | 14 km S of Salton City, CA |

| Time | Latitude | Longitude | Mag | Location |
|---|---|---|---|---|
| 12:20:33.070Z | 37.7505 | -110.678 | 1.94 | 69 km S of Hanksville, Utah |
| 12:21:12.100Z | 33.176 | -115.9751667 | 1.52 | 14 km S of Salton City, CA |
| 12:27:33.920Z | 19.2753333 | -155.4733333 | 1.97 | 7 km N of Pahala, Hawaii |
| 12:27:58.290Z | 21.7055 | 143.0268 | 4.2 | 233 km NW of Farallon de Pajaros, Northern Mariana Islands |
| 12:32:51.000Z | 59.6917 | -153.302 | 2.0 | 83 km W of Anchor Point, Alaska |
| 12:47:52.790Z | 29.6842 | 140.6573 | 4.4 | Izu Islands, Japan region |
| 12:58:29.000Z | 59.8472 | -151.8779 | 1.6 | 8 km NNW of Anchor Point, Alaska |
| 13:04:56.500Z | 36.4973 | -99.0751 | 2.5 | 13 km ENE of Mooreland, Oklahoma |
| 13:24:25.000Z | 63.2956 | -150.3811 | 1.3 | 72 km W of Cantwell, Alaska |
| 13:24:40.020Z | 37.9365 | -112.5178333 | 1.44 | 14 km NNW of Panguitch, Utah |
| 13:40:55.800Z | 19.5233 | -65.4056 | 2.9 | 129 km NNE of Vieques, Puerto Rico |
| 13:44:45.000Z | 62.7685 | -150.7203 | 1.0 | 58 km NNW of Talkeetna, Alaska |
| 13:54:54.200Z | 42.1193333 | -112.1291667 | 1.23 | 12 km SE of Malad City, Idaho |
| 13:59:34.000Z | 60.1147 | -151.9916 | 1.5 | 38 km NNW of Anchor Point, Alaska |
| 14:05:14.000Z | 60.3639 | -152.3244 | 1.6 | 26 km ESE of Redoubt Volcano, Alaska |
| 14:13:16.260Z | 43.734 | -126.695 | 2.88 | off the coast of Oregon |
| 14:23:08.100Z | 33.905 | -117.853 | 1.92 | 4 km NNE of Placentia, CA |
| 14:33:56.000Z | 63.7233 | -149.2295 | 1.1 | 19 km SW of Healy, Alaska |
| 14:34:34.000Z | 60.4919 | -147.4727 | 1.4 | 73 km ESE of Whittier, Alaska |

| | | | | |
|---|---|---|---|---|
| 14:56:05.000Z | 53.4845 | -165.4779 | 1.6 | 75 km SSE of Akutan, Alaska |
| 14:56:27.600Z | 52.5762 | -171.25 | 2.5 | 9 km N of Amukta Island, Alaska |
| 15:03:20.000Z | 62.8124 | -150.2337 | 1.4 | 54 km N of Talkeetna, Alaska |
| 15:03:29.000Z | 63.5564 | -147.5178 | 1.6 | 73 km ENE of Cantwell, Alaska |
| 15:04:47.000Z | 60.9607 | -149.2739 | 1.5 | 38 km WNW of Whittier, Alaska |
| 15:04:47.400Z | 36.52 | -89.5686667 | 1.24 | 8 km SSW of New Madrid, Missouri |
| 15:23:22.235Z | 38.5091 | -118.4115 | 1.5 | 18 km E of Hawthorne, Nevada |
| 15:27:31.060Z | 33.8981667 | -117.8721667 | 1.11 | 3 km N of Placentia, CA |
| 15:29:46.910Z | 19.4855 | -155.2691667 | 1.71 | 6 km NNW of Volcano, Hawaii |
| 15:31:24.100Z | 33.1571667 | -116.5095 | 1.02 | 12 km NE of Julian, CA |
| 15:43:00.990Z | 33.0308333 | -116.423 | 1.17 | 18 km ESE of Julian, CA |
| 15:43:15.370Z | 40.406 | -122.8936667 | 1.62 | 29 km SE of Hayfork, California |
| 15:47:22.000Z | 64.1524 | -162.0719 | 2.1 | 70 km WNW of Unalakleet, Alaska |
| 15:51:58.130Z | 39.0038333 | -122.5958333 | 2.1 | 5 km NNE of Clearlake, California |
| 16:10:47.780Z | 33.0288333 | -116.4276667 | 1.21 | 17 km ESE of Julian, CA |
| 16:13:35.000Z | 60.7795 | -151.6584 | 1.8 | 22 km WNW of Nikiski, Alaska |
| 16:13:40.000Z | 62.4672 | -172.4307 | 2.7 | 150 km SSW of Gambell, Alaska |
| 16:20:35.600Z | 35.7304 | -97.1577 | 3.4 | 8 km NNE of Luther, Oklahoma |
| 16:26:02.836Z | 36.5007 | -114.0056 | 1.0 | 32 km SSE of Bunkerville, Nevada |

| | | | | |
|---|---|---|---|---|
| 16:37:52.000Z | 59.6876 | -153.4481 | 2.7 | 82 km E of Old Iliamna, Alaska |
| 16:50:10.640Z | 38.8135 | -122.8275 | 1.28 | 7 km WNW of The Geysers, California |
| 17:00:47.780Z | 38.8139992 | -122.8271637 | 1.02 | 7 km WNW of The Geysers, California |
| 17:02:03.910Z | -36.302 | -110.4169 | 4.7 | Southern East Pacific Rise |
| 17:15:39.330Z | 1.2153 | 123.7448 | 4.0 | 42 km NNE of Pimpi, Indonesia |
| 17:20:13.170Z | -14.6624 | 166.7117 | 4.6 | 56 km NW of Port-Olry, Vanuatu |
| 17:32:27.770Z | -16.0617 | 167.5172 | 4.5 | 12 km ENE of Lakatoro, Vanuatu |
| 17:35:18.600Z | -7.2878 | 105.8263 | 4.7 | 50 km S of Binuangeun, Indonesia |
| 17:37:54.050Z | 6.7941 | -82.6932 | 4.6 | 138 km S of Punta de Burica, Panama |
| 17:38:17.000Z | 62.0553 | -150.4984 | 2.4 | 36 km WSW of Y, Alaska |
| 17:46:14.630Z | 46.6726667 | -112.4863333 | 2.16 | 29 km WNW of Helena West Side, Montana |
| 17:47:26.000Z | 61.4983 | -151.1495 | 1.3 | 63 km W of Big Lake, Alaska |
| 17:50:37.000Z | 56.6886 | -135.8091 | 2.3 | 50 km SW of Sitka, Alaska |
| 17:52:53.000Z | 61.3447 | -149.8658 | 1.5 | 14 km N of Anchorage, Alaska |
| 18:03:58.000Z | 58.7242 | -150.5348 | 2.0 | 117 km SSE of Homer, Alaska |
| 18:43:29.900Z | 37.9255 | -121.8286667 | 2.72 | 9 km SSW of Antioch, California |
| 19:03:55.700Z | 36.8149 | -97.6196 | 2.8 | 10 km E of Medford, Oklahoma |
| 19:05:12.520Z | 52.4697 | -169.5199 | 3.1 | 68 km SW of Nikolski, Alaska |
| 19:09:43.330Z | -16.2241 | 167.2254 | 4.6 | 23 km WSW of Lakatoro, Vanuatu |

| | | | | |
|---|---|---|---|---|
| 19:15:14.000Z | 53.6146 | -165.3933 | 1.9 | 63 km SSE of Akutan, Alaska |
| 19:38:23.000Z | 62.6802 | -149.4186 | 1.2 | 53 km NE of Talkeetna, Alaska |
| 19:38:30.700Z | 37.9296667 | -121.8198333 | 1.17 | 8 km S of Antioch, California |
| 19:42:39.460Z | 33.3951667 | -116.8748333 | 1.15 | 4 km NNW of Palomar Observatory, CA |
| 19:50:40.300Z | 38.7838333 | -122.7193333 | 1.02 | 3 km ENE of The Geysers, California |
| 19:56:35.000Z | 60.5818 | -151.0448 | 1.4 | 5 km NNE of Ridgeway, Alaska |
| 20:03:12.850Z | 38.7935 | -122.7573333 | 1.57 | 1 km N of The Geysers, California |
| 20:07:18.630Z | -15.0536 | -173.3595 | 4.7 | 107 km NNE of Hihifo, Tonga |
| 20:10:12.710Z | 33.1143 | 140.4294 | 4.6 | 58 km E of Hachijo-jima, Japan |
| 20:17:24.000Z | 62.3146 | -151.1662 | 1.3 | 54 km W of Talkeetna, Alaska |
| 20:19:02.000Z | 56.8977 | -157.826 | 2.4 | 91 km NE of Chignik Lake, Alaska |
| 20:19:53.060Z | 38.8718333 | -122.8168333 | 1.49 | 9 km NW of Cobb, California |
| 20:37:26.970Z | 38.813 | -122.819 | 1.98 | 6 km NW of The Geysers, California |
| 20:43:43.000Z | 56.9659 | -157.8354 | 2.0 | 97 km NE of Chignik Lake, Alaska |
| 20:44:10.000Z | 61.4562 | -150.5787 | 1.8 | 34 km WSW of Big Lake, Alaska |
| 21:00:43.000Z | 59.8505 | -152.9111 | 1.5 | 61 km W of Anchor Point, Alaska |
| 21:04:59.000Z | 62.6916 | -151.2226 | 1.3 | 70 km NW of Talkeetna, Alaska |
| 21:05:05.260Z | 38.8706665 | -122.8191681 | 1.11 | 9 km WNW of Cobb, California |
| 21:17:20.270Z | -18.4088 | -177.8335 | 4.2 | 264 km NNE of Ndoi Island, Fiji |

| | | | | |
|---|---|---|---|---|
| 21:19:34.240Z | 37.7275 | -122.1318333 | 2.92 | 2 km E of San Leandro, California |
| 21:28:40.000Z | 61.349 | -147.2081 | 1.1 | 52 km WNW of Valdez, Alaska |
| 21:29:30.420Z | 37.6322 | 135.5023 | 4.0 | 140 km NW of Hakui, Japan |
| 21:32:26.000Z | 59.8502 | -152.9187 | 2.8 | 62 km W of Anchor Point, Alaska |
| 21:51:10.260Z | 36.3798333 | -120.9778333 | 1.48 | 22 km NE of King City, California |
| 21:56:55.500Z | 35.7243 | -97.1692 | 3.4 | 7 km NNE of Luther, Oklahoma |
| 22:02:22.000Z | 62.9675 | -150.9626 | 1.7 | 84 km NNW of Talkeetna, Alaska |
| 22:06:51.000Z | 52.8783 | -168.3865 | 2.7 | 33 km ESE of Nikolski, Alaska |
| 22:09:02.260Z | -16.2439 | 167.194 | 5.0 | 27 km WSW of Lakatoro, Vanuatu |
| 22:10:30.000Z | 51.7752 | -176.1105 | 1.6 | 39 km ESE of Adak, Alaska |
| 22:11:36.700Z | 18.4988 | -65.8659 | 3.0 | 7 km NNE of Loiza, Puerto Rico |
| 22:21:07.000Z | 62.8616 | -150.7334 | 1.7 | 67 km NNW of Talkeetna, Alaska |
| 22:25:30.000Z | 62.0937 | -150.3875 | 1.8 | 29 km WSW of Y, Alaska |
| 22:31:55.770Z | 55.7911 | 110.3437 | 4.3 | 22 km SE of Kichera, Russia |
| 22:36:13.100Z | 35.7231 | -97.1584 | 2.7 | 7 km NNE of Luther, Oklahoma |
| 22:45:03.000Z | 36.8858 | -97.3182 | 3.0 | 9 km NNW of Blackwell, Oklahoma |
| 22:50:14.900Z | -16.1348 | 166.94 | 4.8 | 48 km W of Norsup, Vanuatu |
| 22:58:11.000Z | 53.5168 | -167.4947 | 1.3 | 74 km WSW of Unalaska, Alaska |
| 23:00:30.420Z | 37.0488333 | -97.732 | 2.05 | 11 km W of Caldwell, Kansas |

| | | | | |
|---|---|---|---|---|
| 23:01:42.010Z | -16.1651 | 167.274 | 4.5 | 15 km SW of Norsup, Vanuatu |
| 23:03:07.710Z | 38.0966667 | -118.7155 | 1.72 | 48 km S of Hawthorne, Nevada |
| 23:04:17.000Z | 64.9846 | -147.3382 | 1.2 | 22 km NNE of Badger, Alaska |
| 23:04:42.000Z | 59.228 | -139.3656 | 1.8 | 41 km SSE of Yakutat, Alaska |
| 23:12:23.590Z | 19.3303333 | -155.1373333 | 2.44 | 15 km SE of Volcano, Hawaii |
| 23:34:52.000Z | 63.9916 | -148.7346 | 1.7 | 18 km NE of Healy, Alaska |
| 23:36:29.560Z | 34.37 | 26.5347 | 4.5 | 91 km SE of Makry Gialos, Greece |
| 23:42:26.210Z | 0.0178 | 125.9166 | 4.4 | 177 km SSE of Bitung, Indonesia |
| 23:57:43.000Z | 61.7016 | -154.1411 | 1.5 | 155 km NNW of Redoubt Volcano, Alaska |

www.ingramcontent.com/pod-product-compliance
Lightning Source LLC
Chambersburg PA
CBHW030817180526
45163CB00003B/1326